I0017689

THE ULTIMATE
BITCOIN
MINING
HANDBOOK

Strategies, Technologies
and Innovations

THE BLOCKCHAIN ACADEMY

The Ultimate Bitcoin Mining Handbook:

Strategies, Technologies, and Innovations

The Ultimate Bitcoin Mining Handbook:
Strategies, Technologies, and Innovations

First Published August 2024

Uprorr Media
Arlington, VA 22209
www.uprorr.com

Copyright © 2024 The Blockchain Academy LLC

All Rights Reserved. No part of this book may be reprinted or reproduced, or utilized in any form or by any electronic, mechanical, or other means now known or hereafter invented, including photocopying and recording, or in any information storage or retrieval system, without permission in writing from the publishers.

Trademark Notice: Product or corporate names may be trademarks or registered trademarks and are used only for identification and explanation without intent to infringe.

A paperback edition is available as
ISBN 979-8-9912798-1-9

A digital edition is available as
ISBN 979-8-9912798-0-2

The Ultimate Bitcoin Mining Handbook:
Strategies, Technologies, and Innovations

Table of Contents

Part III: Advanced Mining Strategies and Considerations..... 141

Preface

Welcome to "The Ultimate Bitcoin Mining Handbook: Strategies, Technologies, and Innovations" This book is designed to be more than just a resource; it is a workbook companion to the Bitcoin Mining Bootcamp, created and offered by The Blockchain Academy Inc. and the Web3 Certification Board Inc. Whether you are a novice looking to dive into the world of cryptocurrency mining or an experienced miner seeking to optimize your operations, this guide aims to equip you with the knowledge and practical skills needed to succeed.

Why This Book?

Bitcoin mining is often seen as a mysterious and complex field, one that requires specialized knowledge and a keen understanding of both technology and economics. This book demystifies the process, breaking it down into manageable and comprehensible parts. By aligning with the curriculum of the Bitcoin Mining Bootcamp, this guide provides a structured learning path that mirrors the comprehensive training provided by our experts at The Blockchain Academy Inc. and the Web3 Certification Board Inc.

A Comprehensive Learning Experience

The chapters in this book cover a wide range of topics, from the fundamental principles of blockchain and cryptocurrency to the intricate details of setting up and optimizing a mining operation. You will find detailed explanations of mining hardware and software, strategies for energy efficiency, and insights into the regulatory and economic aspects of mining. Each section is designed to build upon the last, ensuring a progressive and cohesive learning experience.

Practical Application

One of the key strengths of this book is its focus on practical application. Throughout the chapters, you will find step-by-step guides, troubleshooting tips, and real-world case studies that provide a hands-on approach to learning. Whether you are setting up your first mining rig or looking to scale your operations, the practical insights and exercises included in this guide will help you apply what you learn directly to your mining activities.

Regulatory and Environmental Considerations

In today's rapidly evolving digital landscape, staying informed about regulatory changes and environmental impacts is crucial. This book dedicates significant attention to these areas, providing you with the latest information on compliance, sustainability, and the legal considerations that impact cryptocurrency mining. Understanding these aspects is vital for long-term success and responsible mining practices.

Future-Proofing Your Operations

The world of cryptocurrency mining is dynamic, with constant advancements in technology and market trends. This guide not only addresses current practices but also looks ahead to future developments. By exploring emerging technologies such as AI and IoT, as well as trends in renewable energy, this book prepares you to adapt and thrive in the evolving mining ecosystem.

Collaborative Effort

The creation of this workbook has been a collaborative effort, drawing on the expertise and experience of professionals from The Blockchain Academy Inc. and the Web3 Certification Board Inc.

We are committed to providing the highest quality education and resources to help you succeed in the world of cryptocurrency mining.

Invitation to Learn and Grow

We invite you to use this book as a tool for learning and growth. Whether you are participating in the Bitcoin Mining Bootcamp or using this guide independently, we hope it provides you with valuable insights and practical skills. The journey into Bitcoin mining is filled with opportunities and challenges, and we are here to support you every step of the way.

Thank you for choosing this comprehensive guide as your companion in your mining endeavors. Let's embark on this exciting journey together and explore the vast potential of Bitcoin mining.

The Blockchain Academy LLC

Forward

Foreword to The Ultimate Bitcoin Mining Handbook: Strategies, Technologies, and Innovations

Welcome to *The Ultimate Bitcoin Mining Handbook: Strategies, Technologies, and Innovations*. This comprehensive manual serves as an essential guide for anyone invested in understanding and excelling in the field of Bitcoin mining. As the President of the Texas Blockchain Council, I am honored to share my insights and enthusiasm for this transformative technology and its profound impact on the global economy.

The Genesis of Bitcoin Mining

Bitcoin, introduced by the pseudonymous Satoshi Nakamoto in 2008, heralded a new era of decentralized finance. This revolutionary concept of a peer-to-peer electronic cash system was underpinned by blockchain technology—a secure, transparent, and immutable ledger. At the heart of this system lies Bitcoin mining, the process by which new bitcoins are created and transactions are verified and added to the blockchain.

Bitcoin mining began as a niche activity, primarily conducted by early adopters and tech enthusiasts. Using personal computers, these pioneers mined bitcoins with relatively simple algorithms and minimal computational power. However, as the value and popularity of Bitcoin surged, so did the complexity of the mining process. Today, Bitcoin mining has evolved into a sophisticated and competitive industry, requiring advanced technologies, strategic planning, and significant capital investment.

The Evolution of Bitcoin Mining

The rapid evolution of Bitcoin mining is a testament to human ingenuity and technological progress. What began as a hobbyist activity has transformed into a global industry driven by innovation and competition. This evolution can be traced through several key phases:

1. **CPU Mining**: In the early days, miners used Central Processing Units (CPUs) to solve cryptographic puzzles and earn bitcoins. This method was relatively simple and accessible, but it soon became apparent that CPUs lacked the computational power needed for efficient mining.

2. **GPU Mining**: The next phase saw miners turning to Graphics Processing Units (GPUs), which offered significantly higher processing power. GPU mining quickly became the standard, allowing miners to solve complex puzzles more quickly and earn greater rewards.

3. **FPGA Mining**: Field-Programmable Gate Arrays (FPGAs) brought further advancements, offering even greater efficiency and performance. FPGAs could be programmed to execute specific tasks, making them highly effective for Bitcoin mining.

4. **ASIC Mining**: The most significant leap came with the introduction of Application-Specific Integrated Circuits (ASICs). These specialized chips are designed exclusively for mining, providing unparalleled speed and efficiency. ASIC miners have become the backbone of the Bitcoin mining industry, enabling large-scale operations and significant increases in mining difficulty.

5. **Modern Innovations**: Today, the industry continues to evolve with innovations such as immersion cooling, renewable energy integration, and decentralized mining pools. These advancements not only enhance efficiency but also address environmental concerns and promote sustainability.

The Strategic Landscape of Bitcoin Mining

Successful Bitcoin mining requires more than just technological prowess; it demands strategic acumen and a deep understanding of market dynamics. The ever-changing landscape of Bitcoin mining presents numerous challenges and opportunities, making it imperative for miners to stay informed and adaptable.

1. **Cost Management**: Mining profitability is heavily influenced by operational costs, particularly electricity. Miners must constantly seek ways to optimize energy consumption, reduce costs, and maximize returns. This involves exploring alternative energy sources, negotiating favorable power contracts, and implementing energy-efficient technologies.

2. **Regulatory Compliance**: The regulatory environment for Bitcoin mining varies widely across jurisdictions. Miners must navigate complex legal landscapes, ensuring compliance with local, national, and international regulations. This includes understanding tax implications, licensing requirements, and environmental standards.

3. **Market Trends**: The value of Bitcoin and the difficulty of mining are subject to market fluctuations. Miners must stay abreast of market trends, analyzing factors such as price volatility, network hash rate, and technological advancements. This knowledge enables miners to make informed decisions and adapt their strategies accordingly.

4. **Security and Risk Management**: The decentralized nature of Bitcoin makes it inherently secure, but mining operations are not immune to risks. Miners must implement robust security measures to protect their assets, data, and infrastructure. This includes safeguarding against cyberattacks, physical threats, and operational disruptions.

The Future of Bitcoin Mining

The future of Bitcoin mining is both promising and challenging. As the industry matures, it will continue to face technical, economic, and regulatory hurdles. However, these challenges also present opportunities for innovation and growth.

1. **Sustainability**: One of the most pressing issues is the environmental impact of Bitcoin mining. The industry is increasingly focused on sustainability, with many miners transitioning to renewable energy sources such as solar, wind, and hydroelectric power. Innovations in energy efficiency and carbon offsetting are also gaining traction, paving the way for a more sustainable future.

2. **Decentralization**: The concentration of mining power in a few large operations poses a risk to the decentralized nature of Bitcoin. Efforts to promote decentralization, such as decentralized mining pools and incentives for smaller miners, are crucial for maintaining the integrity and security of the Bitcoin network.

3. **Technological Advancements**: The relentless pace of technological progress will continue to drive the evolution of Bitcoin mining. Advances in hardware, software, and infrastructure will enhance efficiency, reduce costs, and improve scalability. Emerging technologies such as quantum computing may also play a role, presenting both opportunities and challenges for the industry.

4. **Global Adoption**: As Bitcoin gains mainstream acceptance, the demand for mining services will increase. This growth will drive further investment and innovation, fostering a dynamic and competitive industry. Miners will need to adapt to changing market conditions, regulatory frameworks, and technological landscapes to remain successful.

Conclusion

The Ultimate Bitcoin Mining Handbook: Strategies, Technologies, and Innovations is an invaluable resource for anyone involved in or interested in Bitcoin mining. This comprehensive guide covers the full spectrum of mining activities, from the basics of blockchain technology to advanced strategies for optimizing efficiency and profitability.

As you delve into the pages of this handbook, you will gain a deeper understanding of the complexities and opportunities within the Bitcoin mining industry. Whether you are a novice miner seeking to learn the fundamentals or an experienced professional looking to refine your strategies, this manual provides the knowledge and insights you need to succeed.

The future of Bitcoin mining is bright, and the journey is just beginning. I invite you to embrace this exciting world, explore the innovations, and contribute to the ongoing evolution of this transformative technology. Together, we can shape the future of decentralized finance and unlock the full potential of the Bitcoin network.

Lee Bratcher

President, Texas Blockchain Council

The Blockchain Academy LLC

About the Authors

This book is a collaborative effort by a team of dedicated professionals at The Blockchain Academy Inc. and the Web3 Certification Board Inc. Our combined expertise in blockchain technology, cryptocurrency mining, and educational development has allowed us to create a comprehensive and up-to-date guide on Bitcoin mining.

The Blockchain Academy Inc. The Blockchain Academy Inc. is a leading institution committed to advancing the understanding and application of blockchain technology. Our team comprises industry experts, educators, and technologists who are passionate about sharing their knowledge and driving innovation in the blockchain space. With a focus on practical and theoretical education, we aim to equip individuals and organizations with the skills needed to succeed in the rapidly evolving world of blockchain and cryptocurrency.

The Web3 Certification Board Inc. The Web3 Certification Board Inc. is dedicated to setting the standards for excellence in blockchain and Web3 education. Our certification programs are designed to validate the skills and knowledge of professionals in the industry, ensuring they meet the highest standards of competency. Our team includes experienced blockchain practitioners, educators, and certification specialists who work tirelessly to develop and maintain rigorous certification standards.

Our Vision for This Book The creation of this book was driven by our shared belief that Bitcoin mining is a fundamental and enduring aspect of the cryptocurrency ecosystem. As we move further into 2024, the importance of having an up-to-date and authoritative resource on Bitcoin mining has never been greater.

This book is designed to educate and inspire new miners, providing them with the knowledge and tools they need to succeed in this dynamic industry.

We recognize the challenges and opportunities that come with Bitcoin mining, and our goal is to demystify the process, highlight best practices, and offer practical guidance. Through this collaborative effort, we hope to foster a new generation of miners who are informed, skilled, and ready to contribute to the ongoing growth and sustainability of the cryptocurrency mining industry.

Thank you for joining us on this journey into the world of Bitcoin mining. We are excited to share our knowledge and passion with you and look forward to seeing the positive impact of your contributions to this vibrant and evolving field.

Contributors:

- Ryan Mann, Cryptocurrency Mining Expert
- Ryan Williams, Blockchain Educator and Technologist
- Bryant D Nielson, Certification Standards Specialist

We hope you find this book both informative and inspiring as you embark on your own Bitcoin mining adventure.

The Blockchain Academy Inc.

The Web3 Certification Board Inc.

August 2024

Introduction

In the rapidly evolving world of digital currencies, Bitcoin stands as the pioneering cryptocurrency that has fundamentally transformed the way we perceive and engage in financial transactions. As we delve into the intricacies of Bitcoin mining, it is crucial to first understand what Bitcoin is, why it holds such significance, and how mining plays a pivotal role in its ecosystem. This introduction will provide a comprehensive overview of Bitcoin, elucidate the importance of mining, and outline the objectives and structure of this book.

Overview of Bitcoin and Its Significance

The Birth of Bitcoin

Bitcoin was introduced in 2009 by an anonymous entity known as Satoshi Nakamoto, who released the Bitcoin whitepaper titled "Bitcoin: A Peer-to-Peer Electronic Cash System." This whitepaper laid the foundation for a decentralized digital currency that operates without the need for a central authority such as a bank or government. Bitcoin leverages blockchain technology to achieve decentralization, transparency, and immutability.

Key Characteristics of Bitcoin

Decentralization: Unlike traditional currencies, Bitcoin is not controlled by any central authority. Instead, it operates on a peer-to-peer network where transactions are validated by network participants (nodes) through a process known as mining.

Limited Supply: Bitcoin has a finite supply capped at 21 million coins. This scarcity is built into its protocol and is intended to create a deflationary effect, similar to precious metals like gold.

Transparency and Immutability: All Bitcoin transactions are recorded on a public ledger called the blockchain. Once a transaction is added to the blockchain, it cannot be altered or deleted, ensuring transparency and security.

Digital and Borderless: Bitcoin transactions can be conducted globally without the need for intermediaries, making it an efficient means of transferring value across borders.

Security: Bitcoin employs robust cryptographic techniques to secure transactions and control the creation of new units, making it resistant to fraud and counterfeiting.

Bitcoin's Impact and Use Cases

Since its inception, Bitcoin has grown from a niche digital experiment to a global financial phenomenon with a market capitalization that has reached hundreds of billions of dollars. Its impact is far-reaching, influencing various sectors:

Finance: Bitcoin offers an alternative to traditional banking and financial services, enabling users to store and transfer value without relying on intermediaries.

Investment: Bitcoin is often referred to as "digital gold" due to its potential as a store of value and a hedge against inflation. It has attracted significant investment from individuals and institutions alike.

Remittances: Bitcoin facilitates low-cost, efficient cross-border payments, making it an attractive option for remittances, particularly in regions with limited access to banking services.

Decentralized Finance (DeFi): Bitcoin has inspired the development of DeFi applications that aim to recreate traditional financial instruments in a decentralized architecture.

Importance of Mining in the Cryptocurrency Ecosystem

What is Bitcoin Mining?

Bitcoin mining is the process by which new bitcoins are created and transactions are added to the blockchain. Miners use powerful computers to solve complex mathematical problems, which validate transactions and secure the network. This process is known as proof of work (PoW).

Role of Miners

Miners play a crucial role in the Bitcoin network by:

Securing the Network: Miners validate transactions and ensure the integrity of the blockchain. Each block added to the blockchain contains a list of recent transactions that miners have verified as legitimate.

Creating New Bitcoins: Miners are rewarded with newly created bitcoins and transaction fees for their efforts. This is how new bitcoins enter circulation.

Maintaining Decentralization: By distributing the task of validating transactions across a global network of miners, Bitcoin avoids centralization and reduces the risk of a single point of failure.

The Mining Process

The mining process involves several key steps:

1. **Transaction Collection:** Miners collect pending transactions from the network and group them into a block.
2. **Hashing:** Miners then use a cryptographic algorithm (SHA-256) to create a unique hash for the block. This hash must meet a certain difficulty target set by the Bitcoin protocol.
3. **Proof of Work:** Miners compete to find a hash that meets the difficulty target by varying a small part of the block

called the nonce. The first miner to find a valid hash broadcasts the block to the network.

4. **Block Verification:** Other miners and nodes in the network verify the validity of the new block. Once verified, the block is added to the blockchain, and the miner is rewarded with new bitcoins and transaction fees.
5. **Chain Continuation:** The process repeats with miners working on the next block, ensuring the continuous addition of new blocks to the blockchain.

Significance of Mining

Bitcoin mining is essential for several reasons:

Security: Mining ensures that the Bitcoin network remains secure and resistant to attacks. The proof-of-work mechanism makes it computationally expensive to alter transaction history, protecting the network from double-spending and other fraudulent activities.

Incentivization: Mining rewards provide economic incentives for participants to contribute computational power to the network, ensuring its robustness and decentralization.

Supply Regulation: Mining is the mechanism through which new bitcoins are introduced into circulation. The controlled and predictable issuance rate contributes to Bitcoin's scarcity and deflationary nature.

Objectives and Structure of the Book

Objectives

This book aims to provide a comprehensive guide to Bitcoin mining, catering to both beginners and experienced miners. The key objectives are:

Education: To demystify the technical aspects of Bitcoin mining and provide a clear understanding of how the process works.

Practical Guidance: To offer step-by-step instructions for setting up and optimizing mining operations, including hardware and software configuration, energy management, and troubleshooting.

Strategic Insights: To explore advanced mining strategies, economic considerations, and future trends, helping miners maximize profitability and stay ahead in a competitive landscape.

Regulatory Awareness: To highlight the regulatory and legal aspects of Bitcoin mining, ensuring that miners operate within the bounds of the law and understand the risks involved.

Sustainability: To address the environmental impact of mining and discuss sustainable practices and energy-efficient solutions.

Structure

The book is divided into three parts, each focusing on different aspects of Bitcoin mining:

Part I: Fundamentals of Cryptocurrency and Mining

This part provides a solid foundation for understanding the basics of Bitcoin and cryptocurrency mining. It covers the core principles of blockchain technology, the significance of cryptocurrencies, and the fundamental aspects of Bitcoin mining. Readers will learn about the history and evolution of mining hardware, the various types of mineable coins, and the critical components needed to start mining. This section is designed to equip readers with the essential knowledge required to navigate the complexities of the Bitcoin mining ecosystem.

Value of Part I:

- Establishes a strong foundational understanding of blockchain and Bitcoin.
- Clarifies the technical concepts and terminology essential for mining.

- Prepares readers for the more practical aspects of setting up and running a mining operation.

Part II: Setting Up and Optimizing Your Mining Operation

This part delves into the practical aspects of Bitcoin mining. It provides detailed, step-by-step guidance on setting up a mining rig, configuring mining software, and optimizing mining operations for maximum efficiency and profitability. Readers will explore the differences between solo and pool mining, the benefits and challenges of using third-party hosting services, and the various types of energy sources suitable for mining. Additionally, this section addresses important considerations for electrical and network setups, ensuring that readers can build and maintain a robust and efficient mining infrastructure.

Value of Part II:

- Offers practical, actionable steps for setting up and optimizing mining operations.
- Helps readers make informed decisions about hardware, software, and energy sources.
- Provides troubleshooting tips and maintenance advice to keep mining operations running smoothly.

Part III: Advanced Mining Strategies and Considerations

This part focuses on advanced strategies and considerations for experienced miners looking to scale their operations and maximize profitability. It covers topics such as mining hash rate optimization, scaling operations, and future trends in cryptocurrency mining. Readers will gain insights into the economic and legal aspects of mining, including regulatory risks and compliance strategies. Additionally, this section addresses environmental sustainability, discussing energy-efficient practices and the use of renewable energy sources. Case studies and practical workshops provide real-world examples and hands-on experiences to enhance learning.

Value of Part III:

- Provides advanced strategies for optimizing and scaling mining operations.
- Explores the economic and regulatory landscape, helping miners navigate potential risks.
- Emphasizes the importance of sustainability and environmental responsibility in mining practices.

This structure ensures a thorough understanding of Bitcoin mining, guiding readers from the basics to advanced strategies and practical applications. By the end of this book, readers will be equipped with the knowledge and skills to navigate the complex world of Bitcoin mining effectively.

The Blockchain Academy LLC

Part I:
Fundamentals of Cryptocurrency and Mining

The Blockchain Academy LLC

Chapter 1:
What is Blockchain?

Definition and Key Concepts of Blockchain Technology

At its core, a blockchain is a decentralized and distributed digital ledger that records transactions across multiple computers in a way that ensures the security, transparency, and immutability of the data. The term "blockchain" comes from its structure, where individual records, known as "blocks," are linked together in a single continuous "chain." Each block contains a list of transactions and is cryptographically secured, making it virtually impossible to alter once added to the blockchain.

Key concepts of blockchain technology include:

Decentralization: Unlike traditional centralized databases controlled by a single entity, a blockchain operates on a peer-to-peer network of computers (nodes). Each node maintains a copy of the entire blockchain, ensuring that no single point of failure or control exists.

Immutability: Once data is recorded in a block and added to the blockchain, it cannot be altered or deleted. This is achieved through cryptographic hashing and consensus mechanisms, which ensure the integrity and permanence of the recorded data.

Transparency: All transactions recorded on a public blockchain are visible to anyone with access to the blockchain network. This transparency fosters trust and accountability, as all participants can verify the authenticity of the transactions.

Security: Blockchain employs advanced cryptographic techniques to secure data. Each block contains a unique hash (a fixed-length string of characters generated by a cryptographic algorithm) and the hash of the previous block, creating a chain of blocks that is tamper-proof.

Consensus Mechanisms: To add a new block to the blockchain, network participants must reach a consensus. Common consensus mechanisms include Proof of Work (PoW) and Proof of Stake (PoS), which ensure that all nodes agree on the validity of transactions and the state of the blockchain.

How Blockchain Works

To understand how blockchain works, it is essential to delve into the process of creating and adding blocks to the chain, as well as the mechanisms that ensure the security and integrity of the data.

1. **Transaction Initiation:** The process begins when a user initiates a transaction. This transaction could be the transfer of cryptocurrency, the execution of a smart contract, or any other type of data transfer.
2. **Transaction Verification:** Once the transaction is initiated, it is broadcast to the network of nodes. These nodes validate the transaction using predefined rules and protocols. For example, in the case of Bitcoin, nodes verify that the sender has sufficient funds and that the transaction follows the network's rules.
3. **Block Creation:** Validated transactions are grouped together into a block by a node called a "miner." The miner then competes to solve a complex mathematical puzzle, which requires significant computational power. This process is known as "mining" and is the core of the Proof of Work consensus mechanism.
4. **Proof of Work:** The miner who solves the puzzle first broadcasts the block to the network. Other nodes verify the solution and, if it is correct, add the block to their copy of the blockchain. The miner is rewarded with newly created cryptocurrency (e.g., bitcoins) and transaction fees.

5. **Block Addition:** Each block contains a unique hash and the hash of the previous block. This chaining of blocks ensures that altering any block would require recalculating the hashes of all subsequent blocks, which is computationally infeasible.

6. **Consensus Mechanism:** The consensus mechanism ensures that all nodes agree on the current state of the blockchain. In Proof of Stake, for example, validators are chosen to create new blocks based on the number of coins they hold and are willing to "stake" as collateral.

7. **Finality and Immutability:** Once a block is added to the blockchain and confirmed by the network, it becomes a permanent part of the ledger. This immutability ensures that the data cannot be tampered with or deleted.

The Role of Blockchain in the Digital Economy

Blockchain technology has far-reaching implications for the digital economy, offering numerous benefits and applications across various sectors.

Financial Services: Blockchain is revolutionizing the financial industry by enabling faster, cheaper, and more secure transactions. Cryptocurrencies like Bitcoin and Ethereum provide alternatives to traditional banking systems, offering borderless transactions and financial inclusion for unbanked populations. Additionally, blockchain-based smart contracts automate and streamline processes such as insurance claims, trade finance, and asset management.

Supply Chain Management: Blockchain enhances transparency and traceability in supply chains by providing an immutable record of every transaction and movement of goods. This reduces fraud, counterfeiting, and inefficiencies, ensuring that products are sourced, manufactured, and delivered with integrity. For example, blockchain can verify the authenticity of pharmaceuticals, track the origin of food products, and ensure the ethical sourcing of raw materials.

Healthcare: Blockchain improves data security, interoperability, and patient privacy in the healthcare sector. By creating a decentralized and tamper-proof record of patient data, healthcare providers can share information securely and efficiently. This enhances patient care, reduces administrative costs, and mitigates the risk of data breaches. Blockchain also facilitates the tracking of pharmaceuticals and medical devices, ensuring their authenticity and safety.

Digital Identity: Blockchain enables the creation of secure and self-sovereign digital identities, allowing individuals to control and manage their personal information. This reduces the risk of identity theft and fraud, simplifies authentication processes, and enhances privacy. Digital identities can be used for various applications, including online banking, voting, and access to government services.

Voting and Governance: Blockchain enhances the integrity and transparency of voting systems, ensuring that votes are cast, counted, and recorded accurately. This reduces the risk of fraud and manipulation, fostering trust in democratic processes. Blockchain can also be used for decentralized governance models, enabling stakeholders to participate in decision-making processes and ensuring accountability.

Intellectual Property and Copyright: Blockchain provides a transparent and tamper-proof record of intellectual property rights, ensuring that creators receive proper attribution and compensation for their work. This reduces piracy and counterfeiting, incentivizing innovation and creativity. Blockchain can also facilitate the licensing and distribution of digital content, streamlining royalty payments and reducing intermediaries.

Real Estate: Blockchain simplifies and secures real estate transactions by providing a transparent and immutable record of property ownership, transfer, and financing. This reduces fraud, streamlines processes, and lowers transaction costs. Blockchain can also facilitate fractional ownership and crowdfunding of real estate projects, democratizing access to property investments.

Energy and Utilities: Blockchain enables the creation of decentralized energy markets, allowing consumers to buy and sell energy directly from each other. This enhances the efficiency and sustainability of energy distribution, reduces costs, and incentivizes the use of renewable energy sources. Blockchain can also facilitate the tracking and trading of carbon credits, promoting environmental responsibility.

Examples of Blockchain Applications Beyond Cryptocurrency

Smart Contracts: Smart contracts are self-executing contracts with the terms of the agreement directly written into code. These contracts automatically execute when predefined conditions are met, eliminating the need for intermediaries and reducing the risk of human error. For example, a smart contract can automate the release of funds for a freelance project upon the completion and approval of work.

Decentralized Finance (DeFi): DeFi refers to a suite of financial services built on blockchain technology, including lending, borrowing, trading, and insurance. DeFi platforms operate without intermediaries, offering greater accessibility, transparency, and efficiency. For instance, users can lend their cryptocurrency to others and earn interest, or trade digital assets on decentralized exchanges.

Non-Fungible Tokens (NFTs): NFTs are unique digital assets that represent ownership of a specific item, such as art, music, or virtual real estate. NFTs are created and traded on blockchain platforms, ensuring the authenticity and provenance of the digital assets. For example, artists can tokenize their work and sell it directly to collectors, receiving royalties on secondary sales.

Cross-Border Payments: Blockchain enables faster, cheaper, and more secure cross-border payments by eliminating intermediaries and reducing transaction costs. Traditional cross-border payments can take several days and involve multiple intermediaries, each charging fees. Blockchain-based payments, on the other hand, are settled in near real-time and at a fraction of the cost.

Decentralized Autonomous Organizations (DAOs): DAOs are organizations governed by smart contracts and operated by a decentralized community of stakeholders. DAOs enable transparent and democratic decision-making processes, where members can propose, vote on, and implement changes. For example, a DAO can manage a decentralized investment fund, where members pool their resources and vote on investment decisions.

Gaming and Virtual Worlds: Blockchain enhances gaming and virtual worlds by providing secure ownership of in-game assets, enabling players to buy, sell, and trade items across different platforms. Blockchain also facilitates the creation of decentralized gaming economies, where players can earn cryptocurrency by participating in games and contributing to the ecosystem.

Charity and Philanthropy: Blockchain enhances transparency and accountability in charitable organizations by providing a clear and immutable record of donations and their allocation. Donors can track their contributions and ensure that funds are used for their intended purposes. Blockchain can also facilitate cross-border donations, reducing administrative costs and ensuring that more funds reach those in need.

Agriculture and Food Safety: Blockchain improves transparency and traceability in agriculture and food supply chains, ensuring that products are sourced, processed, and delivered with integrity.

Consumers can trace the origin of their food, verify its authenticity, and ensure that it meets safety and quality standards. Blockchain can also facilitate fair trade practices, ensuring that farmers and producers receive fair compensation for their products.

Blockchain technology represents a paradigm shift in how data is recorded, shared, and secured. Its decentralized, transparent, and immutable nature offers numerous benefits and applications beyond cryptocurrency, revolutionizing various sectors of the digital economy. As we continue to explore and innovate, blockchain has the potential to create more efficient, secure, and inclusive systems, driving the future

Chapter 2:
What is Cryptocurrency (From a Mining Perspective)?

Overview of Cryptocurrency

Cryptocurrencies are digital or virtual currencies that leverage cryptographic principles to secure transactions, control the creation of new units, and verify asset transfers. Unlike traditional fiat currencies issued by governments and central banks, cryptocurrencies operate on decentralized networks based on blockchain technology. The most well-known and pioneering cryptocurrency is Bitcoin, introduced in 2009 by an anonymous entity known as Satoshi Nakamoto. Since then, thousands of alternative cryptocurrencies, or "altcoins," have emerged, each with unique features and use cases.

Cryptocurrencies aim to provide an alternative to traditional financial systems, offering benefits such as decentralization, transparency, security, and reduced transaction costs. They enable peer-to-peer transactions without the need for intermediaries like banks, making financial services more accessible and inclusive, especially in regions with limited banking infrastructure.

From a mining perspective, cryptocurrencies are significant because they rely on a process called mining to secure the network, validate transactions, and create new coins. Mining involves solving complex mathematical problems using computational power, which ensures the integrity and immutability of the blockchain.

Nodes

In the context of cryptocurrency networks, nodes are essential components that maintain the network's integrity and functionality.

A node is any computer or device that connects to the cryptocurrency network and participates in its operations. Nodes come in various types, each serving different functions:

Full Nodes: These nodes store a complete copy of the blockchain and validate all transactions and blocks. Full nodes play a crucial role in maintaining the security and decentralization of the network by enforcing the consensus rules. They verify the validity of transactions and blocks, ensuring that no double-spending or invalid data is added to the blockchain.

Lightweight Nodes (SPV Nodes): Lightweight nodes, also known as Simplified Payment Verification (SPV) nodes, do not store the entire blockchain. Instead, they download only the block headers and rely on full nodes to verify transactions. SPV nodes are less resource-intensive and are commonly used in mobile wallets and other applications where storage and processing power are limited.

Mining Nodes: These nodes are operated by miners and are responsible for creating new blocks and adding them to the blockchain. Mining nodes perform the computational work required to solve cryptographic puzzles and validate transactions. Once a mining node successfully mines a new block, it broadcasts the block to the network for validation by other nodes.

Masternodes: Some cryptocurrencies, like Dash, use a specialized type of node called a masternode. Masternodes perform additional functions beyond standard node operations, such as enabling instant transactions, participating in governance decisions, and providing increased privacy features. Masternodes typically require a significant collateral of the cryptocurrency to operate, incentivizing operators to act in the network's best interest.

Nodes communicate with each other to propagate transactions and blocks across the network, ensuring that all participants have a consistent view of the blockchain. This peer-to-peer communication is fundamental to the decentralized nature of cryptocurrencies.

P2P (Peer-to-Peer) Network

A peer-to-peer (P2P) network is a decentralized network architecture where each participant (node) has equal status and can initiate or complete transactions without relying on a central server or intermediary. In the context of cryptocurrencies, P2P networks enable the direct exchange of data between nodes, ensuring that the network remains robust, resilient, and censorship-resistant.

Key characteristics of P2P networks include:

Decentralization: There is no central authority or single point of failure in a P2P network. Each node operates independently and contributes to the overall functionality and security of the network.

Scalability: P2P networks can scale horizontally by adding more nodes, which distribute the workload and increase the network's capacity to handle transactions and data.

Redundancy: Since each node stores a copy of the blockchain or relevant data, P2P networks offer redundancy and fault tolerance. If one or more nodes go offline, the network can continue to operate without disruption.

Security: P2P networks enhance security by distributing data across multiple nodes, making it difficult for attackers to compromise the network. The consensus mechanisms used in P2P networks, such as Proof of Work or Proof of Stake, further protect against malicious activities.

In a cryptocurrency P2P network, when a user initiates a transaction, it is broadcast to all nodes. Nodes verify the transaction according to the network's consensus rules and propagate it further until it reaches miners or validators, who include it in a new block. This decentralized and collaborative process ensures the integrity and security of the blockchain.

A Block

A block is a fundamental unit of data in a blockchain, containing a list of transactions and other relevant information. Each block is linked to the previous block, forming a continuous chain of blocks, hence the term "blockchain." The structure and contents of a block can vary depending on the cryptocurrency and its underlying protocol, but typically include the following components:

1. **Header:** The block header contains metadata about the block, including:
   - **Previous Block Hash:** A reference to the hash of the previous block, linking the blocks together and ensuring the integrity of the blockchain.
   - **Merkle Root:** A hash representing the root of the Merkle tree, which summarizes all transactions in the block.
   - **Timestamp:** The time at which the block was created.
   - **Nonce:** A value used by miners in the Proof of Work process to find a valid hash for the block.
   - **Difficulty Target:** The current difficulty level that determines how hard it is to mine a new block.
2. **Transactions:** The main body of the block contains a list of transactions that have been validated and included by the miner. Each transaction records the transfer of cryptocurrency between addresses and includes details such as the sender, receiver, amount, and transaction fee.
3. **Block Size:** The total size of the block, usually measured in bytes. Different cryptocurrencies have varying block size limits, which can impact the network's capacity and scalability.

The process of creating and adding a block to the blockchain involves several steps:

1. **Transaction Verification:** Miners or validators verify the validity of transactions and group them into a candidate block.
2. **Proof of Work:** Miners compete to solve a cryptographic puzzle by finding a nonce that, when combined with the block's data, produces a hash below the difficulty target. This process requires significant computational power and ensures the security of the network.
3. **Block Propagation:** Once a miner finds a valid hash, the block is broadcast to the network. Other nodes verify the block's validity and, if accepted, add it to their copy of the blockchain.
4. **Chain Continuation:** The process repeats with miners working on the next block, ensuring the continuous addition of new blocks to the blockchain.

Gossip Protocol

The gossip protocol is a communication method used in decentralized networks, including blockchain networks, to efficiently propagate information among nodes. In a gossip protocol, nodes randomly select a subset of their peers and share new information, such as transactions or blocks. These peers then propagate the information to their own peers, and the process continues until the information has reached all nodes in the network.

Key features of the gossip protocol include:

Efficiency: The gossip protocol ensures that information spreads quickly and efficiently throughout the network, even in large and dynamic environments.

Robustness: The protocol is resilient to node failures and network partitions, as the information is redundantly propagated across multiple paths.

Scalability: The gossip protocol can scale to accommodate a large number of nodes without significant performance degradation.

In the context of cryptocurrency networks, the gossip protocol ensures that transactions and blocks are propagated quickly and reliably. When a node receives a new transaction or block, it verifies the information and then shares it with a subset of its peers. This process continues until the information has reached all nodes, ensuring that the network remains consistent and up-to-date.

Ledger

A ledger is a record-keeping system that tracks transactions and balances. In traditional financial systems, ledgers are maintained by centralized entities like banks or accounting firms. In contrast, cryptocurrencies use decentralized ledgers, where transactions are recorded and verified by a network of nodes.

The blockchain itself serves as a decentralized ledger, providing a transparent and immutable record of all transactions. Each block in the blockchain contains a list of transactions, and the entire blockchain represents the complete transaction history of the network.

Key characteristics of the decentralized ledger include:

Transparency: All transactions are publicly visible and can be independently verified by any node in the network. This transparency fosters trust and accountability.

Immutability: Once a transaction is recorded in a block and added to the blockchain, it cannot be altered or deleted. This immutability ensures the integrity and permanence of the data.

Security: The decentralized nature of the ledger, combined with cryptographic techniques and consensus mechanisms, ensures that the data is secure and resistant to tampering.

Decentralization: The ledger is maintained by a distributed network of nodes, eliminating the need for a central authority and reducing the risk of a single point of failure.

Hash

A hash is a fixed-length string of characters generated by a cryptographic algorithm from an input of any size. Hash functions are essential in blockchain technology for ensuring data integrity, security, and immutability. The most commonly used hash function in cryptocurrencies is SHA-256 (Secure Hash Algorithm 256-bit).

Key properties of hash functions include:

Deterministic: The same input will always produce the same hash output.

Fixed Length: Regardless of the input size, the hash output is always a fixed length (e.g., 256 bits for SHA-256).

Pre-image Resistance: It is computationally infeasible to reverse-engineer the original input from its hash output.

Collision Resistance: It is highly unlikely for two different inputs to produce the same hash output.

Avalanche Effect: A small change in the input results in a significantly different hash output.

In the context of blockchain, hashes are used to:

Secure Transactions: Each transaction is hashed, ensuring its integrity and making it tamper-proof.

Link Blocks: Each block contains the hash of the previous block, creating a chain of blocks that is resistant to tampering.

Proof of Work: Miners compete to find a nonce that, when combined with the block's data, produces a hash below the difficulty target. This process secures the network and ensures the integrity of the blockchain.

Blockchain

A blockchain is a decentralized and distributed digital ledger that records transactions across multiple computers in a way that ensures the security, transparency, and immutability of the data. The term "blockchain" comes from its structure, where individual records, known as "blocks," are linked together in a single continuous "chain."

Each block contains a list of transactions, a timestamp, and a cryptographic hash of the previous block. This chaining of blocks ensures that altering any block would require recalculating the hashes of all subsequent blocks, making it computationally infeasible and securing the integrity of the blockchain.

The blockchain operates on a peer-to-peer network, where nodes communicate and share information using the gossip protocol. Consensus mechanisms, such as Proof of Work or Proof of Stake, ensure that all nodes agree on the validity of transactions and the state of the blockchain.

Key features of blockchain technology include:

Decentralization: The blockchain operates on a peer-to-peer network, eliminating the need for a central authority and reducing the risk of a single point of failure.

Immutability: Once a transaction is recorded in a block and added to the blockchain, it cannot be altered or deleted, ensuring the integrity and permanence of the data.

Transparency: All transactions are publicly visible and can be independently verified by any node in the network, fostering trust and accountability.

Security: The blockchain employs advanced cryptographic techniques and consensus mechanisms to secure data and protect against tampering.

Cryptocurrencies represent a revolutionary shift in how we perceive and engage with financial transactions. From a mining perspective, understanding the fundamental concepts of nodes, peer-to-peer networks, blocks, gossip protocols, ledgers, hashes, and blockchain technology is essential. These elements work together to create a decentralized, transparent, and secure digital currency system that offers numerous benefits and applications beyond traditional financial systems.

By leveraging the power of blockchain technology, cryptocurrencies enable peer-to-peer transactions, enhance financial inclusion, and provide a robust and resilient alternative to traditional centralized systems. As we continue to explore and innovate in this space, the potential for cryptocurrencies and blockchain technology to transform various sectors of the digital economy remains vast and promising.

Chapter 3:

Wallets and Keys

Types of Cryptocurrency Wallets

Cryptocurrency wallets are essential tools for managing digital assets. They store the public and private keys needed to access and spend cryptocurrencies and provide an interface for managing balances and transactions. There are several types of wallets, each with its own advantages and disadvantages. The primary types of cryptocurrency wallets include:

Hardware Wallets:

- **Description:** Physical devices designed to securely store private keys offline.
- **Examples:** Ledger Nano S, Trezor, KeepKey.
- **Pros:** High security, immune to online hacking, ideal for long-term storage.
- **Cons:** More expensive than other types, less convenient for frequent transactions.

Software Wallets:

- **Desktop Wallets:**
  - o **Description:** Software applications installed on a personal computer.
  - o **Examples:** Electrum, Exodus, Armory.
  - o **Pros:** More secure than online wallets, provides full control over private keys.
  - o **Cons:** Vulnerable to malware and hacking if the computer is compromised.
- **Mobile Wallets:**
  - o **Description:** Apps installed on a smartphone.

- o **Examples:** Trust Wallet, Mycelium, Coinbase Wallet.
- o **Pros:** Convenient for on-the-go transactions, often include additional features like QR code scanning.
- o **Cons:** Less secure than hardware wallets, susceptible to phone theft and malware.
- **Web Wallets:**
  - o **Description:** Online services accessible through a web browser.
  - o **Examples:** Blockchain.info, Coinbase, MetaMask.
  - o **Pros:** Easily accessible from any device with internet access, user-friendly interfaces.
  - o **Cons:** Less secure as private keys are often stored online, vulnerable to phishing and hacking.

Paper Wallets:

- **Description:** Physical documents that contain printed private and public keys.
- **Pros:** Immune to online hacking, ideal for cold storage.
- **Cons:** Susceptible to physical damage or loss, inconvenient for frequent transactions.

Cold Wallets:

- **Description:** Any wallet that is not connected to the internet, including hardware and paper wallets.
- **Pros:** High security, ideal for long-term storage.
- **Cons:** Less convenient for regular use.

Hot Wallets:

- **Description:** Wallets that are connected to the internet, including web, desktop, and mobile wallets.
- **Pros:** Convenient for frequent transactions, easy to access.
- **Cons:** More vulnerable to hacking and online threats.

Setting Up and Securing Your Wallet

Setting up and securing a cryptocurrency wallet involves several steps to ensure the safety of your digital assets. Here's a detailed guide:

1. **Choosing the Right Wallet:**
   - Consider your needs (e.g., frequency of transactions, security level required).
   - Research different wallet options and read user reviews.
   - Choose a reputable wallet provider with strong security features.
2. **Downloading and Installing the Wallet:**
   - Download the wallet software from the official website or app store.
   - Follow the installation instructions provided by the wallet provider.
   - Verify the authenticity of the software to avoid malware.
3. **Creating a New Wallet:**
   - Open the wallet application and select the option to create a new wallet.
   - Follow the prompts to generate a new wallet, which typically includes creating a strong password and writing down a recovery phrase (seed phrase).
   - Important: Safely store the recovery phrase in a secure location, as it is the only way to recover your wallet if you lose access.
4. **Securing Your Wallet:**
   - Enable two-factor authentication (2FA) if available.
   - Regularly update the wallet software to benefit from security patches and improvements.
   - Use a hardware wallet or cold storage for long-term holdings.
   - Avoid sharing your private keys or recovery phrase with anyone.

- Backup your wallet regularly and store backups in secure, separate locations.
5. **Transferring Funds to Your Wallet:**
   - Obtain your wallet's public address (also known as the receiving address).
   - Use this address to transfer cryptocurrency from an exchange or another wallet to your new wallet.
   - Verify that the transfer is complete by checking your wallet's transaction history and balance.

Private Keys

Private keys are a critical component of cryptocurrency wallets, serving as the key to accessing and managing your digital assets. Understanding and securing private keys is essential for maintaining the integrity and security of your cryptocurrency holdings.

1. **Definition:**
   - A private key is a randomly generated string of characters that allows you to access and control your cryptocurrency. It is used to sign transactions and prove ownership of the assets associated with a specific public key.
2. **Importance of Private Keys:**
   - Private keys are the most crucial piece of information in cryptocurrency management. Possession of the private key gives full control over the corresponding funds.
   - Losing a private key means losing access to the cryptocurrency, as there is no way to recover it without the key.
3. **Generating Private Keys:**
   - Private keys are typically generated by the wallet software during the wallet creation process.
   - They can be generated using various methods, including deterministic algorithms that derive keys from a single seed phrase.

4. **Storing Private Keys:**
   o Store private keys securely, preferably offline in a hardware wallet or a paper wallet.
   o Avoid storing private keys on internet-connected devices or online services.
   o Use strong, unique passwords and encryption to protect digital copies of private keys.
5. **Best Practices for Private Key Security:**
   o Never share your private key with anyone.
   o Regularly backup private keys and store backups in secure, separate locations.
   o Use multisignature wallets, which require multiple private keys to authorize a transaction, for added security.

Public Keys

Public keys are another essential component of cryptocurrency wallets, working in tandem with private keys to enable secure transactions and verify ownership of digital assets.

1. **Definition:**
   o A public key is a cryptographic code that is paired with a private key. It is derived from the private key using a one-way cryptographic function and can be shared publicly.
2. **Role of Public Keys:**
   o Public keys are used to generate public addresses (receiving addresses) for receiving cryptocurrency.
   o They enable others to verify the authenticity of a signed transaction without revealing the private key.
3. **Generating Public Keys:**
   o Public keys are generated from the private key using an elliptic curve cryptography algorithm.
   o The process is one-way, meaning it is computationally infeasible to derive the private key from the public key.

4. Public Key Infrastructure (PKI):

- o The public key infrastructure supports the use of public and private keys, providing a framework for key generation, distribution, and management.
- o PKI ensures the integrity and authenticity of digital certificates and encrypted communications.

Bitcoin Address

A Bitcoin address is a unique identifier that allows users to send and receive Bitcoin. It is derived from the public key and serves as the destination for Bitcoin transactions.

1. Definition:

- o A Bitcoin address is a string of alphanumeric characters that represents a destination for Bitcoin payments.
- o Example: 1A1zP1eP5QGefi2DMPTfTL5SLmv7DivfNa

2. Generating a Bitcoin Address:

- o A Bitcoin address is generated from the public key using a hashing algorithm (SHA-256 followed by RIPEMD-160).
- o The resulting hash is encoded using Base58Check encoding to produce the address.

3. Types of Bitcoin Addresses:

- o **P2PKH (Pay-to-Public-Key-Hash):** The most common type, starting with "1."
- o **P2SH (Pay-to-Script-Hash):** Used for multisignature and other complex scripts, starting with "3."
- o **Bech32 (SegWit):** A newer address format that improves efficiency and security, starting with "bc1."

4. Using Bitcoin Addresses:

- o To receive Bitcoin, share your Bitcoin address with the sender.
- o To send Bitcoin, enter the recipient's Bitcoin address and the amount to be sent in your wallet software.

5. **Security Considerations:**
   - Always verify the Bitcoin address before sending funds to ensure it is correct and not tampered with.
   - Use unique addresses for each transaction to enhance privacy and security.

Account Balance

The account balance in a cryptocurrency wallet represents the total amount of cryptocurrency held by the wallet's public addresses. Managing and tracking your account balance is crucial for effective cryptocurrency management.

1. **Calculating Account Balance:**
   - The account balance is calculated by summing the amounts of all unspent transaction outputs (UTXOs) associated with the wallet's public addresses.
   - UTXOs are the remaining portions of Bitcoin that have not been spent and are available for future transactions.
2. **Viewing Account Balance:**
   - Most wallet applications provide an interface to view your account balance, along with transaction history and other relevant information.
   - The balance is typically displayed in the cryptocurrency's native unit (e.g., BTC for Bitcoin) and may also show the equivalent value in fiat currency.
3. **Transaction History:**
   - The transaction history provides a detailed record of all incoming and outgoing transactions associated with your wallet.
   - Each transaction includes information such as the transaction ID, date, amount, sender, and recipient addresses.
4. **Managing Account Balance:**
   - Regularly monitor your account balance and transaction history to ensure the accuracy and security of your funds.

o Use wallet features such as labels and notes to organize and track your transactions.

5. **Security Best Practices:**
   o Enable notifications for transactions to stay informed about any changes to your account balance.
   o Regularly backup your wallet and private keys to prevent loss of funds due to hardware failure or other issues.
   o Use hardware wallets or cold storage for large balances and long-term holdings to minimize the risk of theft.

Cryptocurrency wallets and keys are fundamental components of the digital asset ecosystem, providing the means to securely store, manage, and transact with cryptocurrencies. Understanding the different types of wallets, the importance of private and public keys, and how to generate and secure Bitcoin addresses is essential for anyone involved in cryptocurrency mining or investing.

By following best practices for wallet setup, security, and management, users can protect their digital assets and ensure the integrity of their transactions. As the cryptocurrency landscape continues to evolve, staying informed about the latest developments and security measures will be crucial for maintaining the safety and usability of cryptocurrency wallets.

Chapter 4:
Bitcoin Mining Basics

Introduction to Mining

Bitcoin mining is a fundamental process that underpins the Bitcoin network, ensuring its security, decentralization, and continuous operation. Mining involves using specialized hardware to solve complex mathematical problems, which validate transactions and add new blocks to the blockchain. This process is crucial for maintaining the integrity of the Bitcoin ledger and preventing double-spending, where the same Bitcoin is used more than once.

Mining is both a competitive and rewarding endeavor. Miners contribute computational power to the network, competing to solve cryptographic puzzles. The first miner to solve the puzzle gets to add a new block to the blockchain and is rewarded with newly created bitcoins and transaction fees. This incentivizes miners to participate in securing the network, making Bitcoin mining a critical component of the cryptocurrency ecosystem.

Proof of Work (PoW) Concept

The Proof of Work (PoW) concept is the backbone of Bitcoin mining. PoW is a consensus mechanism that requires miners to perform computational work to validate transactions and create new blocks. This work involves solving a complex mathematical problem that is difficult to compute but easy to verify. Here's a closer look at how PoW works:

Problem Definition: Miners must find a nonce (a random number) that, when combined with the block's data and passed through a cryptographic hash function (SHA-256 in Bitcoin's case), produces a hash that is lower than a target value set by the network's difficulty.

This target value adjusts periodically to ensure that new blocks are added approximately every 10 minutes.

Computational Effort: Finding the correct nonce requires miners to perform a vast number of hash calculations. This process consumes significant computational power and energy, making it costly and resource-intensive.

Solution Verification: Once a miner finds a nonce that produces a valid hash, they broadcast the new block to the network. Other nodes quickly verify the solution by checking the hash against the target. If the hash is valid, the block is added to the blockchain.

Incentives: The miner who successfully mines a new block is rewarded with newly created bitcoins (block reward) and transaction fees from the transactions included in the block. This reward system incentivizes miners to continue contributing computational power to the network.

Security: PoW ensures the security and integrity of the Bitcoin network. The high computational cost of mining makes it economically impractical for malicious actors to alter transaction history or perform a 51% attack (controlling the majority of the network's hash rate).

Blockchain and the Role of Miners

The blockchain is a decentralized and distributed digital ledger that records all Bitcoin transactions. Miners play a crucial role in maintaining and securing the blockchain. Here's how miners interact with the blockchain:

Transaction Validation: When a Bitcoin transaction is initiated, it is broadcast to the network. Miners collect these unconfirmed transactions and group them into a candidate block. Before adding the block to the blockchain, miners validate the transactions to ensure they are legitimate and adhere to the network's rules.

Block Creation: Miners include the validated transactions in a new block, along with a reference to the previous block (creating a chain of blocks). They then begin the process of solving the PoW puzzle to find a valid nonce.

Securing the Network: By solving the PoW puzzle, miners add new blocks to the blockchain, making it increasingly difficult for anyone to alter previous transactions. Each new block reinforces the security of the previous blocks, creating a robust and tamper-proof ledger.

Decentralization: Miners are distributed across the globe, ensuring that no single entity controls the Bitcoin network. This decentralization enhances the network's resilience to attacks and failures.

Chain Continuation: Once a new block is added to the blockchain, miners start working on the next block, perpetuating the continuous and decentralized operation of the Bitcoin network.

Mining Rewards and Transaction Fees

Mining rewards are the primary incentive for miners to contribute their computational power to the Bitcoin network. These rewards consist of two main components: the block reward and transaction fees.

Block Reward: The block reward is a fixed amount of newly created bitcoins awarded to the miner who successfully mines a new block. Initially set at 50 bitcoins per block, the block reward undergoes a process known as "halving" approximately every four years, reducing the reward by half. This mechanism limits the total supply of bitcoins to 21 million. The most recent halving, in April 2024, reduced the block reward to 3.125 bitcoins per block. The next halving is expected to occur in 2028.

Transaction Fees: In addition to the block reward, miners also earn transaction fees paid by users to prioritize their transactions. When creating a transaction, users can include a fee to incentivize miners to include their transaction in the next block. As the block reward decreases over time, transaction fees are expected to play an increasingly significant role in miners' overall compensation.

The combined reward of newly created bitcoins and transaction fees provides a strong economic incentive for miners to participate in securing the Bitcoin network. As mining becomes more competitive, miners continuously seek ways to optimize their operations, including upgrading hardware, reducing energy costs, and joining mining pools to increase their chances of earning rewards.

What Needs to Be Solved?

Mining is a complex process that involves solving several key challenges to maintain the integrity and security of the Bitcoin network. Here are the main challenges that need to be addressed:

Finding the Nonce: The primary challenge in mining is finding the nonce that, when combined with the block's data, produces a hash that meets the network's difficulty target. This requires miners to perform a vast number of hash calculations, consuming significant computational power and energy.

Balancing Profitability: As the difficulty of mining increases, so does the cost of the computational power and energy required. Miners must balance their operational costs with the potential rewards to ensure profitability. This involves optimizing hardware efficiency, reducing energy consumption, and participating in mining pools to increase the likelihood of earning rewards.

Ensuring Security: Miners must ensure the security of their mining operations to protect against theft, hacking, and other malicious activities. This includes securing their hardware, using robust software, and implementing best practices for network security.

Adapting to Network Changes: The Bitcoin network undergoes periodic adjustments to the difficulty target, block reward halvings, and potential protocol upgrades. Miners must stay informed about these changes and adapt their operations accordingly to remain competitive and profitable.

Sustainability: The energy-intensive nature of PoW mining has raised concerns about its environmental impact. Miners are increasingly seeking sustainable energy sources and more efficient mining techniques to reduce their carbon footprint and ensure the long-term viability of their operations.

Bitcoin mining is a fundamental process that ensures the security, decentralization, and continuous operation of the Bitcoin network. By understanding the basics of mining, including the Proof of Work concept, the role of miners in the blockchain, mining rewards, and the challenges that need to be solved, miners can optimize their operations and contribute to the robust and resilient Bitcoin ecosystem.

As the Bitcoin network continues to evolve, mining will remain a critical component, providing the computational power and security needed to maintain the integrity and trustworthiness of the blockchain. By staying informed about the latest developments and best practices in mining, participants can navigate the complexities of the cryptocurrency landscape and seize the opportunities it offers.

The Blockchain Academy LLC

Chapter 5:
Evolution of Crypto Mining

Historical Development: From CPUs to ASICs

The evolution of cryptocurrency mining is a fascinating journey that mirrors the rapid advancements in technology and the growing popularity of digital currencies. Understanding this evolution is crucial for grasping the current state and future trends of the mining industry.

Early Days: CPU Mining (2009-2010)

When Bitcoin was first introduced by Satoshi Nakamoto in 2009, mining was a relatively simple process that could be performed on ordinary desktop computers using their central processing units (CPUs). In the early days, the Bitcoin network was small, and the computational difficulty of mining was low, making it possible for individuals to mine Bitcoin profitably from their home computers.

- **Simplicity:** CPU mining required minimal technical knowledge and no specialized hardware, making it accessible to a broad audience.
- **Low Competition:** With fewer miners participating, the chances of successfully mining a block and earning the block reward were relatively high.
- **Energy Efficiency:** Compared to later stages, CPU mining consumed less power and generated less heat.

Transition to GPU Mining (2010-2013)

As Bitcoin gained popularity, the network's hash rate increased, leading to a rise in mining difficulty. This prompted miners to seek more powerful solutions. Graphics processing units (GPUs), commonly used in gaming and graphic design, emerged as a more efficient alternative to CPUs.

- **Parallel Processing:** GPUs are designed for parallel processing, allowing them to perform multiple calculations simultaneously. This made them significantly more efficient at solving the cryptographic puzzles required for mining.
- **Increased Hash Rate:** GPU mining offered a substantial increase in hash rate, boosting the chances of successfully mining a block.
- **Adaptability:** Miners could use off-the-shelf graphics cards, making it relatively easy to set up and scale GPU mining rigs.

The Rise of FPGA Mining (2011-2013)

The next significant development in mining hardware was the introduction of field-programmable gate arrays (FPGAs). These devices offered even greater efficiency and performance than GPUs.

- **Customizability:** FPGAs are programmable, allowing miners to optimize the hardware for specific mining algorithms.
- **Energy Efficiency:** FPGAs consume less power per hash compared to GPUs, making them more cost-effective in the long run.
- **Performance:** The ability to configure FPGAs for mining-specific tasks resulted in higher hash rates and faster block processing.

The ASIC Revolution (2013-Present)

The most significant leap in mining hardware came with the advent of application-specific integrated circuits (ASICs). These devices are designed solely for the purpose of mining a particular cryptocurrency, offering unparalleled performance and efficiency.

- **Specialization:** ASICs are tailored for specific mining algorithms, such as Bitcoin's SHA-256, making them extremely efficient at performing the required calculations.
- **High Hash Rate:** ASICs provide a dramatic increase in hash rate compared to CPUs, GPUs, and FPGAs, allowing miners to process blocks more quickly and compete more effectively.
- **Energy Efficiency:** ASICs consume significantly less power per hash than previous mining hardware, reducing operational costs and increasing profitability.
- **Industrial Scale:** The efficiency and performance of ASICs enabled the rise of large-scale mining farms, where thousands of ASIC miners work in tandem to mine cryptocurrency.

The progression from CPUs to ASICs illustrates the relentless pursuit of efficiency and performance in the mining industry. Each technological advancement brought increased hash rates and energy efficiency, but also higher barriers to entry due to the rising cost of hardware and competition.

Technological Advancements in Mining Hardware

The rapid evolution of mining hardware has been driven by the need for greater efficiency and performance. Each stage of development brought new innovations that transformed the mining landscape.

CPU Mining (2009-2010)

- **Single-threaded Performance:** Early Bitcoin miners relied on the single-threaded performance of CPUs to solve cryptographic puzzles. The simplicity and accessibility of CPU mining allowed many individuals to participate.
- **Low Power Consumption:** CPUs consumed relatively low amounts of power, making mining feasible without significant energy costs.

GPU Mining (2010-2013)

- **Parallel Processing Power:** GPUs offered a significant boost in processing power due to their ability to handle parallel tasks. This made them ideal for the repetitive calculations required in mining.
- **Off-the-shelf Availability:** GPUs were widely available and relatively affordable, allowing miners to build powerful rigs by combining multiple graphics cards.
- **Mining Software:** The development of mining software optimized for GPUs, such as CGMiner and BFGMiner, further enhanced the efficiency and effectiveness of GPU mining.

FPGA Mining (2011-2013)

- **Programmability:** FPGAs allowed miners to customize and optimize the hardware for specific mining algorithms. This programmability led to significant gains in efficiency and performance.
- **Energy Efficiency:** FPGAs offered a better balance of power consumption and hash rate compared to GPUs, making them a more cost-effective option for mining.

ASIC Mining (2013-Present)

- **Algorithm-specific Design:** ASICs are designed specifically for mining a particular algorithm, such as Bitcoin's SHA-256. This specialization allows for maximum efficiency and performance.
- **Unmatched Hash Rate:** ASICs provide an exponential increase in hash rate compared to previous hardware, enabling faster block processing and higher rewards.
- **Reduced Power Consumption:** ASICs are highly energy-efficient, consuming less power per hash than CPUs, GPUs,

or FPGAs. This reduces operational costs and increases profitability for miners.

- **Scalability:** The efficiency and performance of ASICs have enabled the development of large-scale mining operations, where thousands of ASIC miners work together in dedicated facilities.

Technological advancements in mining hardware have continuously pushed the boundaries of performance and efficiency. Each innovation has brought new opportunities and challenges, shaping the competitive landscape of the mining industry.

Impacts of Mining Centralization

As mining hardware has evolved, the industry has experienced significant centralization. The transition from CPUs to ASICs has led to the concentration of mining power in the hands of a few large-scale operators. This centralization has several implications for the cryptocurrency ecosystem.

Increased Competition

- Higher Barriers to Entry: The cost of acquiring and operating advanced mining hardware, such as ASICs, has increased significantly. This has raised the barriers to entry, making it difficult for individual miners to compete with large-scale operations.
- **Professionalization:** Mining has evolved from a hobbyist activity to a professional industry. Large-scale mining farms employ specialized teams to manage operations, optimize hardware, and maximize efficiency.

Network Security

- **Enhanced Security:** Large-scale mining operations contribute significant hash power to the network, enhancing its security and resilience against attacks. The more hash power dedicated to mining, the more secure the network becomes.

- **51% Attack Risk:** Despite the increased security, centralization poses the risk of a 51% attack. If a single entity or group of entities controls more than 50% of the network's hash rate, they could potentially manipulate the blockchain, double-spend coins, or prevent transactions from being confirmed.

Decentralization and Distribution

- **Geographic Concentration:** Large mining operations are often concentrated in regions with favorable conditions, such as low electricity costs and cool climates. This geographic concentration can lead to regional centralization of mining power.
- **Regulatory Influence:** The concentration of mining power in specific regions can make the industry vulnerable to regulatory changes. Governments in these regions may impose regulations or restrictions that impact mining operations and the overall network.

Energy Consumption

- **Environmental Impact:** The energy-intensive nature of ASIC mining has raised concerns about its environmental impact. Large-scale mining operations consume significant amounts of electricity, contributing to carbon emissions and environmental degradation.
- **Sustainable Solutions:** In response to environmental concerns, some mining operations are exploring sustainable energy sources, such as hydroelectric, solar, and wind power. These efforts aim to reduce the carbon footprint of mining and promote long-term sustainability.

The centralization of mining power presents both opportunities and challenges for the cryptocurrency ecosystem. While it enhances network security and efficiency, it also raises concerns about decentralization, environmental impact, and regulatory risks.

Choosing the Right Mining Hardware

Selecting the appropriate mining hardware is a critical decision for any miner. The choice of hardware impacts profitability, efficiency, and competitiveness. Several factors should be considered when choosing mining hardware:

Algorithm Compatibility

- **Cryptocurrency Algorithm:** Different cryptocurrencies use different mining algorithms. For example, Bitcoin uses SHA-256, while Ethereum uses Ethash. Ensure that the chosen hardware is compatible with the target cryptocurrency's algorithm.
- **Algorithm Efficiency:** Evaluate the hardware's efficiency for the specific algorithm. ASICs, for instance, are designed for a particular algorithm and offer unmatched efficiency and performance.

Hash Rate and Performance

- **Hash Rate:** The hash rate measures the hardware's ability to solve cryptographic puzzles. Higher hash rates increase the chances of successfully mining a block and earning rewards.
- **Performance Metrics:** Consider other performance metrics, such as power consumption per hash, to evaluate the hardware's overall efficiency. Lower power consumption per hash indicates higher efficiency and profitability.

Cost and ROI

- **Initial Cost:** The initial cost of mining hardware can vary significantly. ASICs tend to be more expensive than GPUs or FPGAs. Assess the budget and determine the maximum investment for hardware.

- **Return on Investment (ROI):** Calculate the potential ROI by considering factors such as hardware cost, hash rate, power consumption, and mining difficulty. Evaluate how long it will take to recover the initial investment and start generating profit.

Energy Efficiency

- **Power Consumption:** Energy costs are a significant factor in mining profitability. Choose hardware that offers a good balance between hash rate and power consumption.
- **Cooling Requirements:** Consider the cooling requirements of the hardware. Efficient cooling solutions can reduce energy costs and extend the lifespan of the hardware.

Longevity and Upgradability

- **Hardware Lifespan:** Assess the expected lifespan of the hardware. ASICs, for example, may become obsolete more quickly than GPUs due to rapid advancements in technology.
- **Upgradability:** Consider the ability to upgrade or repurpose the hardware. GPUs, for instance, can be used for other purposes, such as gaming or AI processing, if they become less profitable for mining.

Manufacturer Reputation

- **Reputation and Support:** Choose hardware from reputable manufacturers with a track record of producing reliable and high-quality products. Consider the availability of customer support and warranty services.
- **Community Feedback:** Research user reviews and community feedback to gain insights into the hardware's performance, reliability, and potential issues.

Environmental Impact

- **Sustainable Practices:** Consider the environmental impact of the chosen hardware. Look for options that support sustainable mining practices, such as energy-efficient designs and the use of renewable energy sources.

By carefully evaluating these factors, miners can select the right hardware to maximize their profitability and competitiveness while minimizing costs and environmental impact.

The evolution of cryptocurrency mining from CPUs to ASICs reflects the relentless pursuit of efficiency and performance in the industry. Each technological advancement brought new opportunities and challenges, shaping the competitive landscape of mining.

The shift from CPU to GPU, FPGA, and ultimately ASIC mining has led to increased hash rates, improved energy efficiency, and the rise of large-scale mining operations. However, it has also contributed to the centralization of mining power, raising concerns about network security, decentralization, and environmental impact.

Choosing the right mining hardware is a critical decision that requires careful consideration of algorithm compatibility, hash rate, performance, cost, energy efficiency, longevity, manufacturer reputation, and environmental impact. By making informed decisions, miners can optimize their operations, contribute to the security and resilience of the cryptocurrency network, and navigate the evolving landscape of the mining industry.

As the cryptocurrency ecosystem continues to grow and evolve, mining will remain a vital component, driving innovation and securing the decentralized future of digital currencies. The lessons learned from the evolution of mining hardware will continue to inform the development of new technologies and practices, ensuring the sustainability and success of the industry.

Chapter 6:
Which Coins to Mine

Overview of Mineable Cryptocurrencies

Cryptocurrency mining is a complex and strategic endeavor, primarily associated with Bitcoin, the first and most well-known cryptocurrency. However, there are numerous other mineable cryptocurrencies, known as altcoins, that present significant opportunities for miners. While this book is primarily focused on Bitcoin, considering the mining of altcoins is crucial for the success and diversification of a crypto miner's portfolio.

Mineable cryptocurrencies use Proof of Work (PoW) or other consensus mechanisms that require computational power to validate transactions and secure the network. Here is an overview of some popular mineable cryptocurrencies:

1. **Bitcoin (BTC):**
   o **Algorithm:** SHA-256
   o **Significance:** The first and most valuable cryptocurrency, Bitcoin is a key player in the digital currency market. It remains the primary focus for many miners due to its high value and widespread adoption.
   o **Mining Difficulty**: High, requiring specialized hardware (ASICs) and significant energy resources.
2. **Litecoin (LTC):**
   o **Algorithm:** Scrypt
   o Significance: Often referred to as the silver to Bitcoin's gold, Litecoin offers faster transaction times and lower fees. It has a strong following and a dedicated mining community.

o **Mining Difficulty**: Moderate, typically mined using ASICs designed for the Scrypt algorithm.

3. **Bitcoin Cash (BCH):**
   o **Algorithm:** SHA-256
   o **Significance**: A fork of Bitcoin aimed at increasing transaction speeds and lowering fees. Bitcoin Cash shares many characteristics with Bitcoin, making it a popular alternative for miners.
   o **Mining Difficulty:** High, also requiring SHA-256 ASICs.

4. **Monero (XMR):**
   o **Algorithm:** RandomX
   o **Significance:** Focused on privacy and anonymity, Monero is designed to be resistant to ASICs, promoting decentralization by enabling CPU and GPU mining.
   o **Mining Difficulty**: Moderate, accessible to individual miners using standard hardware.

5. **Zcash (ZEC):**
   o **Algorithm:** Equihash
   o **Significance:** Known for its strong privacy features, Zcash offers shielded transactions that hide transaction details. It has a dedicated mining community and is typically mined using GPUs.
   o Mining Difficulty: Moderate, with a focus on GPU mining.

6. **Ravencoin (RVN):**
   o **Algorithm:** KawPow
   o **Significance**: Aimed at creating a blockchain optimized for the transfer of assets, Ravencoin uses a mining algorithm designed to be ASIC-resistant, promoting GPU mining.
   o Mining Difficulty: Moderate, ideal for GPU miners.

7. **Dash (DASH):**
   o **Algorithm:** X11

- o **Significance:** Focused on fast, low-cost transactions and decentralized governance, Dash has a unique mining algorithm that combines eleven different hash functions.
- o **Mining Difficulty:** Moderate, typically mined using specialized ASICs.

Additional Mineable Altcoins:

8. **Ethereum Classic (ETC):**
   - o **Algorithm:** Etchash (a modified version of Ethash)
   - o **Significance:** A continuation of the original Ethereum blockchain, Ethereum Classic retains the PoW consensus mechanism and is suitable for GPU mining.
   - o **Mining Difficulty:** Moderate, with a focus on GPU mining.
9. **Grin (GRIN):**
   - o **Algorithm:** Cuckoo Cycle
   - o **Significance:** A privacy-focused cryptocurrency with a unique and scalable blockchain design. Grin is ASIC-resistant, encouraging GPU mining.
   - o **Mining Difficulty:** Moderate, ideal for GPU miners.
10. **Vertcoin (VTC):**
    - o Algorithm: Verthash
    - o Significance: Designed to be ASIC-resistant, Vertcoin promotes decentralized mining using GPUs.
    - o Mining Difficulty: Moderate, accessible to individual miners using standard hardware.
11. **Beam (BEAM):**
    - o **Algorithm:** BeamHashIII
    - o **Significance:** A privacy-focused cryptocurrency using the Mimblewimble protocol. Beam is mined using GPUs and has a strong focus on scalability and security.
    - o **Mining Difficulty:** Moderate, with a focus on GPU mining.

Bitcoin vs. Altcoins

While Bitcoin remains the most well-known and valuable cryptocurrency, mining altcoins can be a strategic and profitable endeavor. Each option has its unique advantages and considerations:

Bitcoin Mining:

- **Pros:**
  - High market value and widespread adoption.
  - Established infrastructure and support.
  - Significant mining rewards despite high difficulty.
- **Cons:**
  - High competition and mining difficulty.
  - Requires significant investment in specialized hardware (ASICs).
  - High energy consumption and operational costs.

Altcoin Mining:

- **Pros:**
  - Potential for higher profitability due to lower mining difficulty.
  - Diversification reduces dependency on a single cryptocurrency.
  - Opportunities to mine with less expensive hardware (GPUs, CPUs).
- **Cons:**
  - Market volatility and lower overall value compared to Bitcoin.
  - Some altcoins may lack liquidity and widespread adoption.
  - Risk of network changes (e.g., Ethereum's transition to PoS).

For a successful mining operation, it is essential to weigh the pros and cons of mining Bitcoin versus altcoins and consider a diversified approach that maximizes profitability and mitigates risks.

Factors to Consider When Choosing a Coin to Mine

Selecting the right cryptocurrency to mine involves evaluating several factors that influence profitability, stability, and long-term viability. Here are the key considerations:

1. **Mining Algorithm:**
   o Different cryptocurrencies use various mining algorithms (e.g., SHA-256, Ethash, Scrypt, RandomX). The choice of algorithm affects the type of hardware required and the overall mining efficiency.
   o Ensure that the chosen algorithm aligns with your available hardware (ASICs, GPUs, CPUs) and expertise.

2. **Mining Difficulty:**
   o Mining difficulty adjusts periodically to maintain a consistent block creation rate. Higher difficulty means more computational power is needed to mine new blocks.
   o Evaluate the current and historical difficulty levels to gauge the feasibility of mining a particular coin.

3. **Block Reward and Halving Events:**
   o The block reward is the number of coins awarded for successfully mining a block. This reward decreases over time through halving events, impacting mining profitability.
   o Consider the current block reward and the schedule of future halvings.

4. **Market Value and Trading Volume:**
   o The market value of a cryptocurrency determines the potential earnings from mining. Higher value coins offer greater rewards, but also attract more competition.

     o Trading volume indicates the liquidity of the coin, affecting the ease of converting mined coins into fiat currency or other assets.

5. **Network Stability and Security:**
   - A stable and secure network ensures reliable mining operations and protects against attacks (e.g., 51% attacks).
   - Research the coin's network security measures and historical performance.

6. **Community and Developer Support:**
   - A strong and active community provides valuable resources, support, and advocacy for the cryptocurrency.
   - Developer involvement indicates ongoing improvements, updates, and innovations, enhancing the coin's long-term prospects.

7. **Energy Consumption and Environmental Impact:**
   - Mining requires significant energy resources, impacting operational costs and environmental sustainability.
   - Evaluate the energy efficiency of the mining hardware and consider using renewable energy sources to mitigate environmental impact.

8. **Regulatory Environment:**
   - Cryptocurrency regulations vary by region and can affect mining operations. Stay informed about local laws and regulations that may impact your chosen coin.

By carefully considering these factors, miners can make informed decisions about which cryptocurrencies to mine, optimizing their operations for profitability and sustainability.

Profitability Comparison of Different Coins

Profitability is a crucial factor in selecting a cryptocurrency to mine. Comparing the potential earnings from different coins involves analyzing various metrics and tools. Here are some key aspects to consider:

1. **Hash Rate and Difficulty:**
   o Calculate the hash rate of your mining hardware for each coin's algorithm.
   o Compare the current mining difficulty to estimate the likelihood of successfully mining new blocks.
2. **Block Reward and Transaction Fees:**
   o Assess the block reward and any additional earnings from transaction fees.
   o Consider how these rewards change over time due to halving events or network adjustments.
3. **Electricity Costs:**
   o Calculate the electricity consumption of your mining hardware and the associated costs.
   o Compare the energy efficiency of different mining setups (e.g., ASICs vs. GPUs) for each coin.
4. **Market Value and Exchange Rates:**
   o Analyze the current market value of each coin and its historical trends.
   o Consider exchange rates and potential fluctuations that may impact the value of mined coins.
5. **Mining Calculators:**
   o Use online mining calculators to estimate potential earnings for different cryptocurrencies. These tools factor in hash rate, difficulty, block reward, electricity costs, and market value.
   o Examples of popular mining calculators include WhatToMine, CoinWarz, and CryptoCompare.
6. **Profitability Analysis:**
   o Perform a profitability analysis by comparing the potential earnings from mining different coins. This involves calculating the net profit after deducting operational costs (e.g., electricity, hardware maintenance).

o Consider the payback period for recovering initial hardware investments and achieving profitability.

7. **Diversification:**
   o Diversifying mining activities across multiple coins can mitigate risks and enhance overall profitability. This strategy leverages the strengths of different cryptocurrencies and adapts to market changes.
   o Monitor market trends and adjust mining operations to capitalize on the most profitable opportunities.

By conducting a thorough profitability comparison, miners can identify the most lucrative cryptocurrencies to mine, optimizing their operations for maximum returns.

Mining cryptocurrencies is a dynamic and multifaceted endeavor that requires careful consideration of various factors. While this book primarily focuses on Bitcoin, exploring the mining of altcoins is essential for achieving a diversified and successful mining operation.

The historical development of mineable cryptocurrencies, from Bitcoin to a wide array of altcoins, highlights the evolution and opportunities in the mining industry. Comparing Bitcoin with altcoins, evaluating key factors when choosing a coin to mine, and conducting a detailed profitability analysis are crucial steps for miners.

By staying informed about market trends, technological advancements, and regulatory changes, miners can navigate the complexities of cryptocurrency mining and optimize their strategies for long-term success. The ability to adapt and diversify mining activities will ensure resilience and profitability in the ever-evolving landscape of digital currencies.

Part II:
Setting Up and Optimizing Your Mining Operation

Chapter 7:
Setting Up Your Mining Rig

Setting up your mining rig is a crucial step in embarking on your Bitcoin mining journey. This chapter provides an in-depth guide on hardware requirements, selecting the right mining rig, building and configuring it, installing mining software, advanced troubleshooting, maintenance techniques, and an overview of the leading mining manufacturers and their presence in the industry.

Hardware Requirements for Mining

To begin mining Bitcoin effectively, you need to invest in specialized hardware designed to handle the demanding computational tasks required. Here's a detailed look at what you'll need:

ASIC Miners: ASIC miners are the backbone of Bitcoin mining. These devices are designed specifically for mining cryptocurrencies and offer unparalleled efficiency and performance.

- **Efficiency:** ASIC miners are optimized for a single task – mining Bitcoin. This specialization makes them far more efficient than general-purpose CPUs or GPUs. Popular models include the Bitmain Antminer and the MicroBT Whatsminer.
- **Hash Rate:** The hash rate, measured in terahashes per second (TH/s), determines how quickly a miner can solve the cryptographic puzzles required to add a block to the blockchain. Higher hash rates increase your chances of earning mining rewards.

Power Supply Unit (PSU): A reliable PSU is essential to power your ASIC miners.

- **Compatibility:** Ensure that your PSU matches the power requirements of your ASIC miner. Check the specifications provided by the manufacturer.
- **Efficiency:** Look for PSUs with an efficiency rating of at least 80 Plus Gold. An efficient PSU reduces electricity consumption, lowering operational costs and environmental impact.

Cooling Solutions: Mining hardware generates significant heat, making effective cooling solutions crucial.

- **Air Cooling:** High-efficiency fans and proper ventilation help prevent overheating. Ensure your setup has good airflow to dissipate heat.
- **Liquid Cooling:** For advanced cooling needs, liquid cooling systems provide superior heat management. They are more effective than air cooling and can prolong the lifespan of your hardware.
- **Immersion Cooling:** The most efficient cooling method involves submerging your hardware in a thermally conductive but electrically insulating liquid. This technique offers optimal cooling for high-performance rigs.

Cables and Connectors: Proper cabling ensures stable power and internet connectivity.

- **Power Cables:** Use the correct power cables to connect your PSU to your ASIC miners securely.
- **Network Cables:** Ethernet cables provide a stable internet connection, crucial for uninterrupted mining operations.

Internet Connection: A stable and fast internet connection is vital for effective mining.

- **Stability:** Ensure that your internet connection is reliable to avoid downtime. Frequent disconnections can lead to lost mining time and reduced profitability.

Monitoring Equipment: Monitoring tools help you keep track of your rig's performance and temperature.

- **Hardware Monitors:** These tools allow you to monitor your mining rig's status, ensuring it operates efficiently and remains within safe temperature ranges.

Selecting the Right Mining Rig

Choosing the right mining rig involves several considerations to ensure you get the best return on your investment:

Budget: Your budget plays a significant role in determining the type of mining hardware you can afford.

- **Initial Investment:** ASIC miners can range from a few hundred to several thousand dollars. Consider what you can comfortably invest upfront.
- **Operational Costs:** Factor in ongoing expenses such as electricity, maintenance, and potential upgrades. Efficient hardware may have higher upfront costs but can save money in the long run through lower energy consumption.

Hash Rate: The performance of your miner, measured by its hash rate, is critical.

- **Performance:** Higher hash rates mean more chances to mine blocks and earn rewards. Strive to balance your budget with the highest hash rate you can afford.

Energy Efficiency: Energy efficiency impacts your operational costs and environmental footprint.

- **Power Consumption:** Compare the power consumption (in watts) to the hash rate to determine efficiency. More efficient miners produce the same hash rate with less power, reducing electricity costs.

Reliability and Support: Choose hardware from reputable manufacturers known for reliability and good customer support.

- **Manufacturer Reputation:** Research the manufacturer's track record and customer reviews.
- **Warranty:** Consider the warranty period and support options. Reliable after-sales support is crucial for addressing any issues that arise.

Resale Value: Future-proof your investment by selecting hardware with good resale value.

- **Future Proofing:** Choose miners that maintain value over time, allowing you to upgrade or exit mining with minimal financial loss.

Step-by-Step Guide to Building and Setting Up Your Mining Rig

Building and setting up your mining rig involves several critical steps:

1. Unboxing and Inspection: Start by carefully unboxing your ASIC miner, PSU, and other components.

- **Unpack Carefully:** Handle components with care to avoid damage.
- **Check Specifications:** Verify that the specifications match what you ordered to ensure compatibility and performance.

2. Assembling the Rig: Begin assembling your mining rig by connecting the PSU to the ASIC miner.

- **Connect PSU to Miner:** Attach power cables securely to prevent power issues.
- **Install Cooling Solutions:** Set up fans or liquid cooling systems. Proper placement and ventilation are essential for efficient cooling.

- **Network Setup:** Connect the miner to your router using an Ethernet cable for a stable internet connection.

3. Initial Setup and Configuration: Power on your rig and access the miner's interface.

- **Power On:** Turn on the PSU and ASIC miner. The miner should boot up automatically.
- **Access Miner Interface:** Use your computer to access the miner's IP address and log in with the default credentials provided by the manufacturer.

4. Configuring Mining Software: Configure the mining software to connect to your chosen mining pool.

- **Mining Pool Configuration:** Enter the mining pool URL, worker ID, and password into the miner's settings.
- **Firmware Updates:** Check for firmware updates from the manufacturer to ensure optimal performance and security.

5. Testing and Calibration: Run the miner to ensure it's operating correctly.

- **Performance Test:** Monitor hash rate, temperature, and power consumption over a few hours.
- **Adjust Settings:** Fine-tune settings based on performance. Adjust fan speeds, clock rates, and voltage as necessary for stability and efficiency.

Installing and Configuring Mining Software

1. Choose Mining Software: Select mining software compatible with your hardware.

- **Software Selection:** Popular options include CGMiner, BFGMiner, and NiceHash. Choose software that supports your hardware and offers desired features.

2. Installation: Download and install the mining software.

- **Download and Install:** Follow installation instructions from the official website. Ensure you download from reputable sources to avoid malware.

3. Configuration: Set up the software with your mining pool details.

- **Pool Information:** Enter the mining pool URL, worker ID, and password.
- **Hardware Configuration:** Adjust settings for your specific hardware, including fan speeds, power limits, and clock speeds.

4. Monitoring and Management: Use monitoring tools to keep track of your rig's performance.

- **Remote Monitoring:** Tools like Awesome Miner and Minerstat offer comprehensive monitoring features.
- **Alerts and Notifications:** Configure alerts for temperature spikes, hash rate drops, or connectivity issues.

Advanced Troubleshooting and Maintenance

Maintaining your mining rig is crucial for sustained performance and profitability.

Common Issues and Solutions: Address common problems to ensure your rig runs smoothly.

- **Overheating:** Ensure proper ventilation and cooling. Regularly clean dust from fans and heat sinks.
- **Network Issues:** Check Ethernet cables and router settings. Ensure your internet connection is stable.
- **Power Failures:** Verify PSU connections and ensure the PSU meets your hardware's power requirements.

Regular Maintenance: Keep your rig in top condition with regular maintenance.

- **Cleaning:** Dust and debris can cause overheating. Clean your miner and cooling systems regularly.
- **Firmware Updates:** Keep your mining firmware up to date for performance improvements and security patches.
- **Performance Checks:** Regularly monitor hash rate, temperature, and power consumption. Adjust settings as needed to maintain optimal performance.

Advanced Diagnostics: Use diagnostic tools to troubleshoot and resolve issues.

- **Log Analysis:** Check mining logs for error messages or performance issues.
- **Hardware Diagnostics:** Use diagnostic tools to test hardware components. Some miners include built-in diagnostic functions.

Top 5 Mining Manufacturers and Their Industry Presence

Choosing the right mining hardware is crucial for successful mining operations. Here are the top five mining hardware manufacturers and their estimated market share:

Bitmain: Bitmain is the largest and most well-known manufacturer of ASIC miners, holding a dominant position in the market.

- **Market Share:** Approximately 60%
- **Popular Models:** Antminer S19 Pro, Antminer S19j Pro, Antminer S9
- **Overview:** Bitmain's Antminer series is renowned for its high efficiency and reliability. Bitmain continues to lead the market with cutting-edge technology and robust customer support.

MicroBT: MicroBT has gained significant market share with its efficient and high-performance miners.

- **Market Share:** Approximately 25%
- **Popular Models:** Whatsminer M30S++, Whatsminer M31S+
- **Overview:** The Whatsminer series from MicroBT is known for high hash rates and energy efficiency. MicroBT has established itself as a strong competitor to Bitmain, providing high-quality miners that are widely adopted in the industry.

Canaan Creative: Canaan Creative is one of the pioneers in the ASIC mining hardware market, known for its AvalonMiner series.

- **Market Share:** Approximately 10%
- **Popular Models:** AvalonMiner 1246, AvalonMiner 1166 Pro
- **Overview:** Canaan Creative's AvalonMiner series is recognized for its stability and performance. Their miners are a popular choice among large-scale mining operations.

Ebang: Ebang focuses on producing efficient and reliable ASIC miners, contributing to its growing market presence.

- **Market Share:** Approximately 3%
- **Popular Models:** Ebit E12+, Ebit E11++
- **Overview:** Ebang's Ebit series offers competitive hash rates and energy efficiency. Ebang continues to innovate and expand its presence in the mining industry.

Innosilicon: Innosilicon provides a range of ASIC miners for different cryptocurrencies, known for their robust performance and efficiency.

- **Market Share:** Approximately 2%
- **Popular Models:** T3+ Pro, T2T-32T

- **Overview:** Innosilicon's Bitcoin miners, such as the T3+ Pro, are known for their high efficiency and performance. Innosilicon continues to innovate and expand its market share.

Setting up a mining rig requires careful planning, the right hardware selection, and precise configuration. By following this comprehensive guide, you can build and maintain a robust mining rig capable of efficiently mining Bitcoin and other cryptocurrencies. Regular maintenance and monitoring will ensure your rig operates at peak performance, maximizing your mining rewards and ensuring a successful mining venture. The top mining manufacturers listed here offer reliable and efficient hardware, essential for staying competitive in the ever-evolving world of cryptocurrency mining.

Chapter 8:
Mining Software & Configuration

Mining software plays a crucial role in the performance and efficiency of your mining operation. This chapter provides an overview of popular mining software, a step-by-step guide for installation and setup, tips for configuring the software for optimal performance, strategies for monitoring and managing your operations, and advanced troubleshooting techniques.

Overview of Popular Mining Software

Selecting the right mining software is essential for optimizing your mining rig's performance. Here are some of the most popular mining software options available:

CGMiner

- **Overview:** CGMiner is one of the oldest and most widely used mining software applications. It supports a wide range of ASIC miners and GPUs.
- **Features:** Open-source, cross-platform (Windows, Linux, macOS), extensive monitoring and control options, remote interface capabilities, and support for multiple mining pools.
- **Best For:** Advanced users who need robust features and control over their mining operations.

BFGMiner

- **Overview:** BFGMiner is similar to CGMiner but is more focused on mining with ASICs and FPGA devices.
- **Features:** Modular design, remote interface capabilities, integrated overclocking and monitoring, fan control, and support for multiple mining algorithms.

- **Best For:** Users with ASIC and FPGA miners looking for detailed customization options.

NiceHash

- **Overview:** NiceHash offers a unique approach by allowing users to sell their hash power and earn Bitcoin. It supports both GPU and ASIC mining.
- **Features:** Easy setup, user-friendly interface, real-time monitoring, automated mining, and profit-switching based on market conditions.
- **Best For:** Beginners and users who want a simple, automated mining experience.

EasyMiner

- **Overview:** EasyMiner is a graphical frontend for mining Bitcoin, Litecoin, and other cryptocurrencies. It supports both solo and pooled mining.
- **Features:** User-friendly interface, real-time statistics, automatic detection of network devices, and configuration files for miners like CGMiner and BFGMiner.
- **Best For:** Users looking for an easy-to-use graphical interface for their mining software.

Awesome Miner

- **Overview:** Awesome Miner is a powerful mining management software that supports over 50 mining engines, including CGMiner, BFGMiner, and SGMiner.
- **Features:** Centralized management for multiple miners, real-time monitoring, profit-switching, and extensive support for different mining hardware.
- **Best For:** Users managing multiple mining rigs and looking for comprehensive monitoring and management tools.

Step-by-Step Installation and Setup

Here's a step-by-step guide to installing and setting up mining software. We'll use CGMiner as an example, but the process is similar for other software:

1. Download the Software:

- **Official Website:** Visit the official website of the mining software to download the latest version. For CGMiner, visit CGMiner GitHub.
- **Download:** Download the appropriate version for your operating system (Windows, Linux, macOS).

2. Install the Software:

- **Unzip Files:** Extract the downloaded files to a directory on your computer.
- **Install Dependencies:** Ensure that all necessary dependencies are installed. For CGMiner, this might include installing libcurl and other libraries.

3. Configure the Software:

Create a Config File: Create a configuration file (usually a .conf or .json file) with your mining settings. Here's an example for CGMiner:

Json

```
{

"pools": [

{

"url": "stratum+tcp://pool_address:port",

"user": "username.worker",
```

```
"pass": "password"

}

],

"api-listen": true,

"api-allow": "W:127.0.0.1",

"log": "5",

"logfile": "/path/to/logfile.log"

}
```

Edit Config File: Replace pool_address, port, username.worker, and password with your mining pool details.

4. Start Mining:

- **Run the Software:** Open a terminal or command prompt, navigate to the directory where you extracted the files, and run the executable file. For CGMiner, you might use:

Sh

./cgminer --config /path/to/config_file.conf

5. Monitor Performance:

- **Check Output:** Monitor the terminal or command prompt for output, ensuring your miner connects to the pool and begins hashing.
- **Adjust Settings:** If needed, adjust your configuration file for optimal performance and restart the software.

Configuring Software for Optimal Performance

Configuring your mining software for optimal performance involves several key settings and adjustments:

Pool Settings:

- **Primary and Backup Pools:** Configure multiple mining pools to ensure continuous mining if the primary pool goes down.
- **Stratum Protocol:** Use the stratum protocol for lower latency and better performance.

Hardware Settings:

- **Hash Rate:** Monitor and adjust the hash rate settings to maximize performance without overheating your hardware.
- **Fan Speed:** Adjust fan speeds to maintain optimal temperatures. Many mining software options allow for automated fan control based on temperature thresholds.
- **Voltage and Frequency:** Fine-tune voltage and frequency settings (overclocking and undervolting) to balance performance and energy efficiency.

Software Settings:

- **Thread Concurrency:** For GPU mining, adjust thread concurrency settings to match your GPU's capabilities.
- **Work Size:** Adjust work size settings for optimal performance based on your hardware.
- **API Access:** Enable API access for remote monitoring and control.

Performance Monitoring:

- **Real-Time Statistics:** Use software features to monitor hash rate, temperature, fan speed, and power consumption in real-time.

- **Logs and Reports:** Enable logging to track performance over time and identify any issues that need attention.

Monitoring and Managing Your Mining Operations

Effective monitoring and management of your mining operations are crucial for maintaining optimal performance and profitability:

Remote Monitoring:

- **Monitoring Tools:** Use tools like Awesome Miner, Minerstat, or native software features to monitor your rigs remotely. These tools provide dashboards with real-time statistics and alerts.
- **Mobile Access:** Many monitoring tools offer mobile apps, allowing you to monitor your rigs from anywhere.

Performance Alerts:

- **Threshold Alerts:** Set up alerts for temperature, hash rate drops, or connectivity issues. These alerts help you respond quickly to potential problems.
- **Automated Actions:** Configure automated actions, such as restarting a miner if it crashes or reconnecting to a different pool if the primary pool is down.

Reporting and Analytics:

- **Detailed Reports:** Generate detailed reports on performance, profitability, and uptime. These reports help you analyze trends and make informed decisions.
- **Profitability Calculations:** Use tools to calculate profitability based on current electricity costs, hash rates, and market prices.

Security Management:

- **Firewall and VPN:** Implement firewall and VPN solutions to protect your mining network from unauthorized access.
- **Regular Updates:** Keep your mining software and firmware updated to protect against vulnerabilities and improve performance.

Advanced Troubleshooting Techniques

When issues arise, advanced troubleshooting techniques can help resolve problems and maintain smooth operations:

Diagnosing Common Issues:

- **Overheating:** Ensure proper ventilation and cooling. Clean dust from fans and heat sinks regularly.
- **Network Issues:** Verify Ethernet cable connections and router settings. Ensure your internet connection is stable and reliable.
- **Power Failures:** Check PSU connections and ensure the PSU is adequately rated for your hardware. Monitor power consumption to avoid overloads.

Log Analysis:

- **Error Logs:** Review mining software logs for error messages or performance issues. Logs often provide insights into the root cause of problems.
- **System Logs:** Check operating system logs for hardware-related issues or crashes.

Hardware Diagnostics:

- **Built-In Tools:** Use diagnostic tools provided by the miner's manufacturer to test hardware components. These tools can identify failing parts or performance bottlenecks.

- **Third-Party Tools:** Utilize third-party diagnostic software to test and benchmark your hardware. Tools like HWMonitor or GPU-Z can provide valuable insights.

Network Troubleshooting:

- **Ping and Traceroute:** Use network diagnostic tools like ping and traceroute to identify network latency or connectivity issues.
- **IP Configuration:** Verify that your mining rig has a static IP address to avoid connectivity problems with the mining pool.

Community Support:

- **Forums and Groups:** Engage with online forums and social media groups dedicated to mining. The mining community can provide valuable advice and support.
- **Manufacturer Support:** Contact the hardware manufacturer's support team for assistance with hardware-specific issues.

Mining software and its configuration are pivotal to the success of your mining operations. By choosing the right software, setting it up correctly, and continuously monitoring and managing your mining rigs, you can maximize performance and profitability. Advanced troubleshooting techniques ensure that you can quickly resolve issues and maintain smooth operations. With the right tools and knowledge, you can optimize your mining setup and achieve success in the competitive world of Bitcoin mining.

Mining Software and its Interaction with Mining Pools

When choosing mining software, it's essential to understand how each program interacts with mining pools, as this can impact your mining efficiency and overall profitability. Here's a detailed comparison of how different mining software programs handle joining and managing mining pools:

1. CGMiner

Overview: CGMiner is a versatile and powerful mining software that supports various mining pools and hardware. It is known for its extensive configuration options and robust performance monitoring.

Joining a Mining Pool:

- **Configuration:** You can join a mining pool by specifying the pool URL, worker name, and password in the configuration file or via command line arguments.
- **Multi-Pool Support:** CGMiner allows you to configure multiple pools, prioritizing them based on your preferences. If the primary pool goes down, it automatically switches to backup pools.
- **Stratum Protocol:** CGMiner supports the Stratum protocol, which reduces network latency and improves mining efficiency.
- **API Access:** It provides API access for remote monitoring and management, allowing you to manage pool settings and monitor performance in real-time.

BFGMiner

Overview: BFGMiner is similar to CGMiner but focuses more on mining with ASIC and FPGA devices. It offers modular design and flexibility for advanced users.

Joining a Mining Pool:

- **Configuration:** Similar to CGMiner, BFGMiner uses configuration files or command line arguments to connect to mining pools.

- **Multi-Pool Management:** BFGMiner supports multiple pools with failover capabilities. You can set priorities for your pools, ensuring continuous mining.
- **Stratum Support:** The software supports Stratum and Getwork protocols, offering flexibility in pool connections.
- **Advanced Features:** BFGMiner includes features like dynamic clocking, monitoring, and remote interface capabilities, allowing for detailed control over pool connections and performance.

NiceHash

Overview: NiceHash offers a unique approach by allowing users to sell their hashing power to buyers who need it. It supports both GPU and ASIC mining.

Joining a Mining Pool:

- **Automated Pool Selection:** NiceHash automatically connects to the most profitable pool based on real-time market conditions. This feature simplifies the process for beginners and ensures maximum profitability.
- **Profit Switching:** The software dynamically switches between different algorithms and pools to maximize earnings.
- **User-Friendly Interface:** NiceHash's interface makes it easy to start mining with minimal configuration. Users can simply input their Bitcoin address and start mining.
- **Limited Control:** While NiceHash offers ease of use, it provides less control over specific pool settings compared to CGMiner and BFGMiner.

EasyMiner

Overview: EasyMiner is a graphical frontend for mining Bitcoin, Litecoin, and other cryptocurrencies. It supports both solo and pooled mining.

Joining a Mining Pool:

- **Graphical Interface:** EasyMiner's GUI simplifies the process of joining a mining pool. Users can input pool details through the interface without dealing with command lines.
- **Pool Configuration:** You can easily configure pool URL, worker name, and password through the software's settings menu.
- **Multi-Pool Support:** While it supports multiple pools, EasyMiner is primarily designed for ease of use, making it suitable for beginners.
- **Integrated Wallet:** EasyMiner integrates a wallet for direct deposit of mining rewards, simplifying the overall mining process.

Awesome Miner

Overview: Awesome Miner is a powerful mining management software that supports over 50 mining engines, including CGMiner and BFGMiner. It is designed for managing large mining operations.

Joining a Mining Pool:

- **Centralized Management:** Awesome Miner provides a centralized dashboard to manage multiple mining rigs and pool connections from a single interface.
- **Multi-Pool Configuration:** It supports configuring multiple pools for failover and load balancing. Users can set priorities and automate pool switching.
- **Profit Switching:** Awesome Miner includes profit-switching capabilities, automatically selecting the most profitable pool and algorithm based on real-time data.
- **Advanced Monitoring:** The software offers comprehensive monitoring tools, including performance statistics, real-time data, and remote access, providing detailed control over pool connections.

Key Differences in Pool Management

Ease of Use:

- **NiceHash and EasyMiner** are designed for ease of use, making them ideal for beginners. They simplify the process of joining and managing pools with user-friendly interfaces and automated features.
- **CGMiner and BFGMiner** offer more control and flexibility, suitable for advanced users who need detailed configuration options and performance monitoring.

Multi-Pool Support:

- **CGMiner, BFGMiner, and Awesome Miner** provide robust multi-pool support with automatic failover and prioritization, ensuring continuous mining even if a pool goes down.
- **NiceHash** offers automated profit switching, connecting users to the most profitable pool dynamically.
- **EasyMiner** supports multiple pools but with fewer advanced options compared to CGMiner and BFGMiner.

Customization and Control:

- **CGMiner and BFGMiner** offer extensive customization options, allowing users to tweak settings for optimal performance and control over pool connections.
- **NiceHash and EasyMiner** focus on simplicity and automation, providing less control but greater ease of use.
- **Awesome Miner** combines ease of use with advanced features, making it suitable for managing large-scale mining operations with detailed monitoring and pool management capabilities.

Profit Optimization:

- **NiceHash and Awesome Miner** include profit-switching capabilities, automatically optimizing mining operations for maximum profitability.
- **CGMiner and BFGMiner** require manual configuration for pool switching and optimization, offering more control but requiring more expertise.

In summary, the choice of mining software depends on your experience level, the scale of your mining operation, and your need for control and customization. Beginners may prefer NiceHash or EasyMiner for their simplicity, while advanced users and large-scale operators may opt for CGMiner, BFGMiner, or Awesome Miner for their extensive features and control over pool management.

The Blockchain Academy LLC

Chapter 9:
Solo vs. Pool Mining

Differences Between Solo and Pool Mining

Cryptocurrency mining involves two primary approaches: solo mining and pool mining. Both methods have unique characteristics, benefits, and challenges, and the choice between them depends on various factors, including individual preferences, resources, and goals.

Solo Mining:

In solo mining, an individual miner uses their own hardware and resources to mine cryptocurrency independently. The miner is responsible for all aspects of the mining process, including validating transactions, solving cryptographic puzzles, and adding new blocks to the blockchain. If successful, the miner receives the entire block reward and transaction fees.

Pool Mining:

Pool mining involves multiple miners combining their computational power to work together as a single entity, known as a mining pool. By pooling resources, miners increase their chances of solving cryptographic puzzles and earning rewards. When a mining pool successfully mines a block, the rewards are distributed among the pool members based on their contributed computational power.

Advantages and Disadvantages of Each Approach

Both solo mining and pool mining have distinct advantages and disadvantages. Understanding these can help miners make informed decisions based on their circumstances and objectives.

Solo Mining:

Advantages:

- **Full Reward:** Solo miners receive the entire block reward and transaction fees for successfully mining a block, maximizing their earnings.
- **Independence:** Solo miners have complete control over their mining operations, hardware, and strategies without needing to coordinate with other miners.
- **No Pool Fees:** Solo miners do not pay any fees to a mining pool, keeping all their earnings.

Disadvantages:

- **High Variance:** Solo mining has a high variance, meaning the rewards can be unpredictable and inconsistent. It may take a long time to successfully mine a block, leading to periods with no earnings.
- **Resource Intensive:** Solo mining requires significant computational power and energy resources. Miners need to invest in high-performance hardware and incur higher operational costs.
- **Difficulty:** As the network difficulty increases, solo miners find it harder to compete with large mining operations and pools, reducing their chances of successfully mining blocks.

Pool Mining:

Advantages:

- **Steady Income:** Pool mining provides a more stable and predictable income stream by distributing rewards among pool members based on their contributed computational power.

- **Lower Variance:** By pooling resources, miners reduce the variance in earnings, receiving smaller but more frequent payouts.
- **Community Support:** Mining pools often provide support, resources, and a community for miners to share knowledge and collaborate on troubleshooting and optimization.

Disadvantages:

- **Pool Fees:** Mining pools charge fees for their services, which are deducted from the rewards. These fees can vary but typically range from 1% to 3%.
- **Centralization:** Pool mining can contribute to centralization in the network if a few large pools dominate the mining power, potentially impacting the network's security and decentralization.
- **Dependence on Pool:** Pool miners rely on the mining pool's infrastructure, management, and policies. Any issues or changes within the pool can affect the miners' earnings and operations.

How to Join a Mining Pool

Joining a mining pool is a straightforward process, but it requires careful consideration and setup to ensure optimal performance and profitability. Here are the steps to join a mining pool:

Research and Choose a Mining Pool:

- Research various mining pools and compare their features, fees, payout structures, and reputations. Consider factors such as pool size, geographic distribution, and community support.
- Choose a pool that aligns with your mining goals, hardware, and preferences.

Create an Account:

- o Visit the mining pool's website and create an account. This typically involves providing an email address and setting up a password.
- o Some pools may require additional verification steps, such as two-factor authentication (2FA) for added security.

Configure Your Mining Software:

- o Download and install mining software compatible with your hardware and the chosen pool. Popular mining software includes CGMiner, BFGMiner, and EasyMiner.
- o Configure the mining software with the pool's server address, port number, and your account credentials. This information is usually provided on the pool's website.

Set Up Your Mining Hardware:

- o Ensure your mining hardware (ASICs, GPUs, CPUs) is correctly set up and connected to the mining software.
- o Optimize your hardware settings for maximum performance and efficiency, including overclocking and cooling solutions.

Start Mining:

- o Launch the mining software and start mining. The software will connect to the pool and begin contributing computational power to the collective mining effort.
- o Monitor your mining performance, hashrate, and earnings through the pool's dashboard and your mining software.

Withdraw Your Earnings:

- o Once you accumulate earnings in the pool, you can withdraw them to your cryptocurrency wallet. Follow the pool's withdrawal procedures and ensure your wallet address is correctly configured.

Popular Mining Pools and Their Features

There are numerous mining pools available, each with its unique features, fees, and payout structures. Here are some of the most popular mining pools and their key characteristics:

Antpool:

- o **Algorithm:** SHA-256 (Bitcoin), Scrypt (Litecoin), and others.
- o **Features:** Antpool is one of the largest mining pools, offering a user-friendly interface, detailed statistics, and multiple payout methods (PPS, PPLNS).
- o **Fees:** 1% to 2.5%, depending on the payout method.
- o **Reputation:** Operated by Bitmain, a leading mining hardware manufacturer, Antpool is well-regarded for its reliability and support.

Slush Pool:

- o **Algorithm:** SHA-256 (Bitcoin), Equihash (Zcash).
- o **Features:** Slush Pool was the first mining pool and remains popular for its transparency, advanced security features, and user-friendly interface.
- o **Fees:** 2% for Bitcoin mining, 0% for Zcash mining.
- o **Reputation:** Known for its transparency and detailed reporting, Slush Pool has a strong reputation in the mining community.

F2Pool:

- o **Algorithm:** SHA-256 (Bitcoin), Scrypt (Litecoin), Ethash (Ethereum Classic), and others.
- o **Features:** F2Pool supports a wide range of cryptocurrencies, offering daily payouts, a mobile app, and detailed mining statistics.
- o **Fees:** 2.5% for Bitcoin mining, 3% for other coins.

o **Reputation:** One of the largest and most diverse mining pools, F2Pool is known for its reliability and comprehensive support for various coins.

BTC.com:

o **Algorithm:** SHA-256 (Bitcoin), Scrypt (Litecoin), and others.
o **Features:** BTC.com provides a user-friendly interface, detailed analytics, and multiple payout methods (FPPS, PPS).
o **Fees:** 1.5% to 4%, depending on the payout method.
o **Reputation:** Operated by Bitmain, BTC.com is highly regarded for its performance, transparency, and robust infrastructure.

ViaBTC:

o **Algorithm:** SHA-256 (Bitcoin), Scrypt (Litecoin), and others.
o **Features:** ViaBTC offers a wide range of cryptocurrencies, flexible payout methods (PPS, PPLNS, SOLO), and advanced monitoring tools.
o **Fees:** 2% for Bitcoin mining, 3% for other coins.
o **Reputation:** Known for its versatility and support for multiple coins, ViaBTC is a popular choice among miners.

SparkPool:

o **Algorithm:** Ethash (Ethereum Classic), Cuckoo Cycle (Grin), and others.
o **Features:** SparkPool focuses on GPU-mined coins, offering low fees, stable payouts, and an easy-to-use interface.
o **Fees:** 1% to 1.5%, depending on the coin.
o **Reputation:** Well-regarded for its performance and low fees, SparkPool is a top choice for GPU miners.

Choosing between solo mining and pool mining is a critical decision for cryptocurrency miners. Each approach has its unique advantages and disadvantages, and the best choice depends on individual goals, resources, and risk tolerance.

Solo mining offers the potential for higher rewards and independence but comes with higher variance and resource requirements. Pool mining provides a more stable income and community support but involves sharing rewards and relying on the pool's infrastructure.

By understanding the differences between solo and pool mining, evaluating the pros and cons of each approach, and carefully selecting a mining pool, miners can optimize their operations and maximize their profitability. Popular mining pools like Antpool, Slush Pool, F2Pool, BTC.com, ViaBTC, and SparkPool offer a range of features and support, making it easier for miners to find a pool that aligns with their needs and preferences.

As the cryptocurrency mining landscape continues to evolve, staying informed about the latest trends, technologies, and best practices will help miners navigate the complexities of the industry and achieve long-term success.

Chapter 10:
Using Mining Facilities with 3rd Party Hosting Services

Introduction to 3rd Party Mining Facilities

As cryptocurrency mining becomes more competitive and resource-intensive, many miners are turning to third-party mining facilities, also known as hosted mining services, to manage their operations. These services allow miners to outsource the setup, maintenance, and management of mining hardware to specialized facilities that offer infrastructure, technical support, and optimized environments for mining.

Third-party mining facilities provide an appealing solution for those who want to mine cryptocurrencies without the hassle of handling hardware and dealing with the complexities of mining setups. These facilities typically offer a range of services, including hardware procurement, installation, maintenance, and sometimes even power and cooling solutions tailored to mining operations.

Pros and Cons of Using Hosted Mining Services

Utilizing third-party mining facilities has both advantages and disadvantages. Understanding these can help miners make informed decisions about whether hosted mining is the right choice for their needs.

Pros:

1. **Reduced Overhead and Maintenance:**
   o **Simplified Setup:** Third-party facilities handle the procurement, installation, and configuration of mining hardware, reducing the complexity for miners.

o **Ongoing Maintenance:** These services provide continuous maintenance and troubleshooting, ensuring that mining hardware operates efficiently and minimizing downtime.

2. **Optimized Environment:**
   o **Cooling Solutions:** Professional mining facilities are equipped with advanced cooling systems designed to maintain optimal temperatures for mining hardware, enhancing performance and longevity.
   o **Power Management:** These facilities often have access to stable and cost-effective power sources, reducing energy costs and ensuring a consistent power supply.

3. **Scalability:**
   o **Ease of Expansion:** Hosted mining services allow miners to scale their operations easily by adding more hardware without worrying about infrastructure constraints.
   o **Flexible Contracts:** Many services offer flexible contracts, allowing miners to adjust their operations based on market conditions and profitability.

4. **Location Advantages:**
   o **Geographic Benefits:** Some mining facilities are located in regions with favorable conditions, such as low electricity costs and cool climates, which can significantly enhance mining efficiency.
   o **Regulatory Considerations:** Facilities in regions with favorable regulatory environments can help mitigate legal risks and uncertainties.

5. **Expertise and Support:**
   o **Professional Management:** Hosted mining services employ experts who manage and optimize mining operations, ensuring high efficiency and profitability.

o **Technical Support:** Access to round-the-clock technical support can quickly address any issues, minimizing downtime and maximizing mining returns.

Cons:

1. **Cost:**
   o **Service Fees:** Hosted mining services charge fees for their services, which can include setup fees, maintenance fees, and a percentage of mining profits. These fees can impact overall profitability.
   o **Hidden Costs:** Some services may have hidden costs, such as additional charges for electricity or upgrades, which can add up over time.
2. **Control and Dependence:**
   o Lack of Control: By outsourcing mining operations, miners relinquish control over their hardware and the operational decisions made by the facility.
   o **Dependence on Provider:** Miners are dependent on the reliability and performance of the hosting service. Any issues with the provider, such as financial instability or poor management, can negatively affect mining operations.
3. **Security Risks:**
   o Data and Asset Security: Storing mining hardware and data in a third-party facility introduces security risks, including potential theft, hacking, or loss of access to the hardware.
   o **Regulatory Risks:** Changes in regulations affecting the hosting provider's location can impact mining operations and profitability.
4. **Transparency:**
   o **Lack of Visibility:** Miners may have limited visibility into the day-to-day operations and performance of their

hardware, making it difficult to verify the accuracy of reports and earnings.

o **Trust Issues:** Trusting a third-party provider with valuable mining hardware and data requires confidence in the provider's integrity and reliability.

How to Choose a Reliable Hosting Service

Selecting a reliable third-party mining facility is crucial for maximizing the benefits of hosted mining while minimizing risks. Here are key factors to consider when choosing a hosting service:

1. **Reputation and Track Record:**
   o **Research Providers:** Investigate the reputation and track record of potential hosting services. Look for reviews, testimonials, and case studies from other miners.
   o **Industry Experience:** Choose providers with a proven history in the mining industry and a track record of successful operations.
2. **Service Offerings and Flexibility:**
   o **Range of Services:** Ensure the provider offers a comprehensive range of services, including hardware procurement, installation, maintenance, and technical support.
   o **Contract Flexibility:** Look for providers that offer flexible contract terms, allowing you to scale operations up or down based on market conditions and profitability.
3. **Infrastructure and Location:**
   o **Facility Quality:** Assess the quality of the hosting facility, including cooling systems, power management, and security measures.
   o **Geographic Advantages:** Consider the location of the facility, taking into account factors such as electricity costs, climate, and regulatory environment.

4. **Transparency and Reporting:**
   o **Detailed Reporting:** Choose a provider that offers transparent and detailed reporting on mining performance, hardware status, and earnings.
   o **Visibility:** Ensure the provider allows for real-time monitoring of mining operations, giving you visibility into the performance and efficiency of your hardware.
5. **Security Measures:**
   o **Physical Security:** Evaluate the physical security measures in place to protect mining hardware from theft and damage.
   o **Cybersecurity:** Assess the provider's cybersecurity protocols to protect data and prevent unauthorized access to mining operations.
6. **Support and Communication:**
   o **Technical Support:** Ensure the provider offers round-the-clock technical support to quickly address any issues that may arise.
   o **Communication:** Look for providers that maintain open and clear communication channels, keeping you informed about the status of your mining operations.
7. **Cost Structure:**
   o **Fee Transparency:** Understand the provider's fee structure, including setup fees, maintenance fees, and profit-sharing arrangements. Be aware of any potential hidden costs.
   o **Cost-Benefit Analysis:** Perform a cost-benefit analysis to determine whether the fees charged by the hosting service are justified by the potential benefits and profitability.
8. **Regulatory Compliance:**
   o **Legal Standing:** Ensure the provider operates in compliance with local regulations and has the necessary licenses and permits to conduct mining operations.

o **Risk Mitigation:** Choose a provider that proactively addresses regulatory risks and adapts to changes in the legal environment.

CASE STUDIES AND EXAMPLES

Case Study 1: BitRiver - Leveraging Renewable Energy

Overview: BitRiver is one of the largest third-party mining facilities in Russia, strategically located in Bratsk, a city known for its cold climate and access to cheap hydroelectric power. The facility utilizes renewable energy to power its mining operations, making it an environmentally friendly option for miners.

Challenges:

- High initial setup costs for infrastructure.
- Ensuring consistent power supply and maintaining optimal hardware performance.

Solutions and Benefits:

- **Renewable Energy:** By leveraging hydroelectric power, BitRiver significantly reduces its operational costs and environmental impact.
- **Cooling Efficiency:** The cold climate of Bratsk provides natural cooling, enhancing hardware efficiency and longevity.
- **Scalability:** BitRiver offers flexible contracts and scalable solutions, allowing miners to easily expand their operations.
- **Security and Support:** The facility provides robust security measures and round-the-clock technical support, ensuring high operational uptime.

Impact: Miners using BitRiver benefit from lower electricity costs, reduced environmental impact, and high operational efficiency, resulting in improved profitability.

Case Study 2: Genesis Mining - Global Reach and Expertise

Overview: Genesis Mining, based in Iceland, is one of the most well-known and reputable third-party mining facilities in the world. The facility takes advantage of Iceland's geothermal and hydroelectric energy sources, offering sustainable and cost-effective mining solutions.

Challenges:

- Managing a diverse client base with varying needs and expectations.
- Maintaining high levels of transparency and customer trust.

Solutions and Benefits:

- **Green Energy:** Genesis Mining utilizes Iceland's renewable energy resources, minimizing its carbon footprint and energy costs.
- **Transparency:** The company provides detailed reporting and real-time monitoring, ensuring transparency and trust with clients.
- **Global Reach:** Genesis Mining's extensive experience and global presence allow it to offer tailored solutions for miners from different regions.
- **Expert Support:** The facility's team of experts provides comprehensive support and optimization services, maximizing mining performance.

Impact: Genesis Mining's commitment to sustainability, transparency, and customer support has established it as a trusted partner for miners seeking reliable and efficient hosted mining solutions.

Case Study 3: Hut 8 Mining - High-Performance Data Centers

Overview: Hut 8 Mining, based in Canada, operates high-performance data centers specifically designed for cryptocurrency mining. The company focuses on providing secure, scalable, and efficient mining environments.

Challenges:

- Ensuring data center security and reliability.
- Managing energy consumption and operational costs.

Solutions and Benefits:

- **High-Performance Data Centers:** Hut 8 Mining's facilities are equipped with state-of-the-art cooling and power management systems, ensuring optimal hardware performance.
- **Security:** The company implements stringent physical and cybersecurity measures to protect mining hardware and data.
- **Energy Efficiency:** Hut 8 Mining leverages Canada's cold climate and access to affordable electricity to reduce energy consumption and costs.
- **Scalability:** The company offers flexible solutions that allow miners to scale their operations as needed.

Impact: Miners using Hut 8 Mining's facilities benefit from secure, efficient, and cost-effective mining environments, leading to enhanced profitability and operational stability.

Case Study 4: Core Scientific - Advanced Technology and Innovation

Overview: Core Scientific, based in the United States, is a leader in blockchain and AI infrastructure. The company offers advanced mining facilities that incorporate cutting-edge technology and innovative solutions to optimize mining performance.

Challenges:

- Staying ahead of technological advancements and market trends.

- Managing large-scale operations and ensuring high uptime.

Solutions and Benefits:

- **Advanced Technology:** Core Scientific uses the latest mining hardware and software, ensuring high efficiency and performance.
- **Innovation:** The company continuously invests in research and development to stay at the forefront of the industry.
- **Uptime Guarantee:** Core Scientific offers a high uptime guarantee, ensuring consistent mining operations and maximizing earnings.
- **Comprehensive Support:** The company provides end-to-end support, from hardware setup to ongoing maintenance and optimization.

Impact: Core Scientific's focus on technology and innovation provides miners with access to state-of-the-art facilities and support, resulting in high-performance and profitable mining operations.

Using third-party mining facilities offers a viable solution for miners seeking to optimize their operations without the complexities and challenges of managing hardware independently. By leveraging the expertise, infrastructure, and support provided by hosting services, miners can enhance the efficiency and profitability of their mining activities.

However, selecting a reliable hosting service requires careful consideration of various factors, including reputation, service offerings, infrastructure, transparency, security, support, cost structure, and regulatory compliance. By thoroughly evaluating potential providers and understanding the pros and cons of hosted mining, miners can make informed decisions that align with their goals and maximize their success in the competitive world of cryptocurrency mining.

As the mining industry continues to evolve, hosted mining services will likely play an increasingly important role, offering innovative solutions and opportunities for miners to thrive in the ever-changing landscape of digital currencies.

Chapter 11:
Electrical and Network Setup

Electrical Basics for Mining

Setting up an efficient and reliable electrical system is crucial for successful cryptocurrency mining. The power requirements for mining operations are substantial, and improper electrical setup can lead to hardware failures, inefficiencies, and safety hazards. Here are the key components and considerations for setting up the electrical infrastructure for mining.

Power Supply Units (PSUs):

- **Selection:** Choose high-quality PSUs that are compatible with your mining hardware. PSUs should have sufficient wattage to handle the combined power draw of all components in your mining rig.
- **Efficiency Rating:** Look for PSUs with high efficiency ratings (80 PLUS Gold or higher) to minimize energy loss and reduce electricity costs.

Circuit Breakers and Wiring:

- **Load Capacity:** Ensure that your electrical circuits can handle the load of your mining equipment. Overloading circuits can cause breakers to trip and pose fire hazards.
- **Dedicated Circuits:** Use dedicated circuits for your mining rigs to avoid overloading household circuits. This helps distribute the electrical load evenly and reduces the risk of electrical failures.

Voltage Stability:

- **Voltage Regulators:** Consider using voltage regulators or uninterruptible power supplies (UPS) to protect your mining equipment from voltage fluctuations and power surges.
- **Consistent Supply:** Ensure a consistent and stable power supply to maintain optimal performance and prevent hardware damage.

Cooling and Ventilation:

- **Heat Management:** Mining hardware generates significant heat, requiring effective cooling solutions. Install adequate ventilation and cooling systems to maintain safe operating temperatures.
- **Airflow:** Optimize airflow within the mining area to prevent overheating and ensure that hot air is expelled efficiently.

Network Configuration

A robust and secure network setup is essential for maintaining the reliability and efficiency of your mining operations. Proper network configuration helps ensure uninterrupted connectivity, reduces latency, and enhances security.

Internet Connection:

- **Bandwidth:** Ensure that you have a high-speed internet connection with sufficient bandwidth to handle the data requirements of your mining operations.
- **Stability:** Choose a reliable internet service provider (ISP) with minimal downtime to maintain continuous mining activities.

Local Area Network (LAN):

- **Wired vs. Wireless:** Use wired connections (Ethernet) for your mining rigs to ensure stable and low-latency connectivity. Wireless connections are more prone to interference and signal loss.

- **Switches and Routers:** Invest in high-quality network switches and routers to manage data traffic efficiently. Ensure that these devices can handle the volume of data generated by your mining operations.

IP Address Configuration:

- **Static IPs:** Assign static IP addresses to your mining rigs for easier management and monitoring. This allows for consistent and predictable network configuration.
- **Subnetting:** Use subnetting to organize and manage your network more effectively, especially if you have multiple mining rigs.

Remote Monitoring and Management

Remote monitoring and management tools are essential for maintaining the efficiency and security of your mining operations. These tools allow you to monitor performance, manage hardware, and respond to issues from anywhere in the world.

Monitoring Software:

- **Real-Time Monitoring:** Use monitoring software that provides real-time data on hashrate, temperature, power consumption, and other key performance metrics.
- **Alerts and Notifications:** Set up alerts and notifications for critical events, such as hardware failures, temperature spikes, and network issues.

Management Platforms:

- **Centralized Management:** Use centralized management platforms to control and configure multiple mining rigs from a single interface. This simplifies administration and enhances operational efficiency.

- **Automation:** Implement automation tools to perform routine tasks, such as rebooting rigs, adjusting settings, and managing updates, reducing the need for manual intervention.

Remote Access:

- **Secure Access:** Ensure that remote access to your mining operations is secure. Use virtual private networks (VPNs), secure shell (SSH) protocols, and strong passwords to protect against unauthorized access.
- **Multi-Factor Authentication (MFA):** Implement MFA for added security when accessing your mining rigs remotely.

Network Protocols and Security

Ensuring the security of your mining network is paramount to protect against cyber threats, unauthorized access, and data breaches. Implementing robust security measures and following best practices can safeguard your mining operations.

Firewall Configuration:

- **Perimeter Security:** Set up firewalls to protect your network perimeter. Configure firewall rules to allow only necessary traffic and block potential threats.
- **Internal Segmentation:** Use internal firewalls to segment your network, isolating critical components and reducing the risk of lateral movement by attackers.

Intrusion Detection and Prevention:

- **IDS/IPS Systems:** Implement intrusion detection systems (IDS) and intrusion prevention systems (IPS) to monitor network traffic and detect malicious activities.
- **Regular Updates:** Keep your IDS/IPS systems updated with the latest threat signatures and patches.

Data Encryption:

- Encrypt Data: Use encryption protocols (e.g., SSL/TLS) to secure data transmitted over the network. This prevents eavesdropping and data interception.
- **VPNs:** Use VPNs to encrypt traffic between remote locations and your mining network, ensuring secure communication.

Access Control:

- **User Authentication:** Implement strong user authentication methods, including MFA, to control access to your mining network and hardware.
- **Role-Based Access:** Use role-based access control (RBAC) to assign permissions based on user roles, minimizing the risk of unauthorized access.

Power Solutions and Backup Planning

Reliable power solutions and backup plans are essential for maintaining continuous mining operations and minimizing downtime. Power outages and disruptions can significantly impact profitability and hardware longevity.

Power Redundancy:

- **Multiple Power Sources:** Use multiple power sources to ensure redundancy. This can include grid power, generators, and renewable energy sources.
- **Automatic Transfer Switches (ATS):** Install ATS to automatically switch to backup power sources in the event of a primary power failure.

Uninterruptible Power Supplies (UPS):

- **Battery Backup:** Use UPS systems to provide temporary battery backup during power outages. This allows you to safely shut down mining rigs and prevent data loss or hardware damage.

- **Surge Protection:** Ensure that UPS systems offer surge protection to guard against voltage spikes and power surges.

Energy Management:

- **Load Balancing:** Implement load balancing techniques to distribute power evenly across your mining rigs, reducing the risk of overloading circuits.
- **Power Monitoring:** Use power monitoring tools to track energy consumption and identify inefficiencies. This helps optimize power usage and reduce costs.

Disaster Recovery Plan:

- **Contingency Planning:** Develop a disaster recovery plan that outlines procedures for dealing with power outages, hardware failures, and other emergencies.
- **Regular Testing:** Regularly test your backup power systems and disaster recovery plan to ensure they function correctly and effectively.

Setting up a robust electrical and network infrastructure is critical for the success and efficiency of cryptocurrency mining operations. By understanding and implementing best practices in electrical setup, network configuration, remote monitoring, network security, and power management, miners can optimize their operations, minimize downtime, and protect their investments.

As the mining industry evolves, staying informed about the latest technologies and practices in electrical and network setup will be essential for maintaining competitive and profitable mining operations. By prioritizing reliability, security, and efficiency, miners can ensure the long-term success and sustainability of their mining endeavors.

Chapter 12:
Types of Energy Used in Bitcoin Mining

Overview of Energy Sources

The energy consumption of Bitcoin mining is a significant factor influencing both operational costs and environmental impact. Understanding the different types of energy sources available for mining can help miners optimize their operations for cost efficiency and sustainability. This chapter explores the primary energy sources used in Bitcoin mining, highlighting their benefits, challenges, and practical applications.

Grid Electricity

Grid electricity is the most commonly used energy source for Bitcoin mining due to its widespread availability and ease of access.

Availability and Cost Considerations:

- **Widespread Availability:** Grid electricity is accessible in most regions, making it a convenient choice for mining operations.
- **Cost Variability:** Electricity costs can vary significantly depending on the location, time of day, and demand. Miners need to consider these factors when calculating profitability.

Environmental Impact:

- **Energy Mix:** The environmental impact of grid electricity depends on the energy mix of the grid. Regions relying heavily on fossil fuels have a higher carbon footprint compared to those utilizing renewable sources.

- **Carbon Emissions:** Mining operations in areas with high fossil fuel usage contribute significantly to carbon emissions and climate change.

Dependability and Grid Stability:

- **Reliability:** Modern electrical grids provide a stable and reliable power supply, minimizing downtime for mining operations.
- **Grid Stability:** High demand from mining operations can strain local grids, potentially leading to instability and power outages.

Renewable Energy

Renewable energy sources, such as solar, wind, and hydro, offer sustainable alternatives to traditional grid electricity. These sources are becoming increasingly popular among miners looking to reduce their environmental impact and long-term costs.

Solar Energy: Setup, Cost, and Efficiency:

- **Setup:** Solar panels can be installed on-site, converting sunlight into electricity. The setup requires significant initial investment in solar panels, inverters, and storage batteries.
- **Cost:** After the initial investment, solar energy provides a low-cost, long-term energy solution. Maintenance costs are relatively low compared to other energy sources.
- **Efficiency:** Solar energy efficiency depends on geographic location and weather conditions. Regions with high sunlight exposure are ideal for solar mining operations.

Wind Energy: Benefits and Challenges:

- **Benefits:** Wind energy is a renewable source with a low environmental footprint. It can generate significant amounts of electricity, especially in regions with consistent wind patterns.

- **Challenges:** Wind turbines require substantial initial investment and are site-specific, needing locations with reliable wind conditions. Visual and noise impacts on local communities can also be a concern.

Hydro Energy: Sustainability and Scalability:

- **Sustainability:** Hydroelectric power is a stable and reliable renewable energy source with a low carbon footprint. It harnesses the energy of flowing water to generate electricity.
- **Scalability:** Hydropower plants can be scaled to meet the energy demands of large mining operations. However, they are limited to regions with suitable water resources.

Case Studies of Successful Renewable Energy Integration:

- **BitRiver:** Located in Bratsk, Russia, BitRiver uses hydroelectric power to run its mining operations. The cold climate and renewable energy source provide an efficient and environmentally friendly solution.
- **Genesis Mining:** Based in Iceland, Genesis Mining leverages the country's abundant geothermal and hydroelectric energy to power its mining facilities, minimizing its carbon footprint and reducing energy costs.

Natural Gas

Natural gas is a fossil fuel that offers a transitional energy solution for Bitcoin mining, providing lower emissions compared to coal and oil.

Cost-Effectiveness and Setup:

- **Cost-Effective:** Natural gas is often more affordable than other fossil fuels, offering a cost-effective energy solution for mining operations.

- **Setup:** Setting up natural gas infrastructure requires pipelines, storage facilities, and combustion systems, which can be complex and costly.

Environmental Considerations:

- **Lower Emissions:** Natural gas combustion produces fewer greenhouse gases and pollutants compared to coal and oil, making it a cleaner fossil fuel alternative.
- **Carbon Footprint:** Despite being cleaner than other fossil fuels, natural gas still contributes to carbon emissions and climate change.

Practical Examples:

- **Layer1:** A Bitcoin mining company in West Texas, Layer1 uses natural gas flaring—capturing natural gas that would otherwise be burned off and wasted—to power its mining operations, reducing environmental impact and utilizing an otherwise lost energy source.

Nuclear Power

Nuclear power provides a high-efficiency, low-carbon energy source suitable for large-scale mining operations.

High Energy Efficiency:

- **Energy Density:** Nuclear power plants produce substantial amounts of electricity from a relatively small amount of fuel, making them highly efficient for energy-intensive activities like mining.
- **Consistent Supply:** Nuclear power provides a stable and continuous electricity supply, unaffected by weather conditions or any time of day.

Safety and Regulatory Considerations:

- **Safety Concerns:** Nuclear power poses safety risks, including potential for catastrophic accidents and challenges in radioactive waste disposal.
- **Regulatory Challenges:** Nuclear energy is heavily regulated, with stringent safety and environmental standards. Obtaining necessary approvals and permits can be time-consuming and complex.

Feasibility and Public Perception:

- **High Initial Costs:** Building and maintaining nuclear power plants require significant financial investment, making it feasible primarily for large-scale mining operations.
- **Public Perception:** Public concerns about safety and environmental impact can affect the acceptance and expansion of nuclear power.

Geothermal Energy

Geothermal energy harnesses heat from the Earth's interior to generate electricity, offering a renewable and sustainable energy source for mining.

Sustainable Energy Production:

- **Renewable Resource:** Geothermal energy is a renewable resource with a minimal environmental footprint, providing a consistent and reliable power supply.
- **Low Emissions:** Geothermal power plants produce low levels of greenhouse gases compared to fossil fuels.

Geographic Limitations:

- **Site-Specific:** Geothermal energy is limited to regions with significant geothermal activity, such as volcanic areas and tectonic plate boundaries.

- **High Initial Investment:** The development of geothermal power plants requires substantial upfront investment in drilling and infrastructure.

Practical Applications in Mining:

- **Iceland's Geothermal Power:** Iceland's abundant geothermal resources power multiple mining operations, including Genesis Mining, providing a sustainable and cost-effective energy solution.
- **Puna Geothermal Venture:** In Hawaii, the Puna Geothermal Venture supplies geothermal energy to local mining operations, demonstrating the practical application of geothermal power in mining.

Considerations for Energy Selection

Selecting the right energy source for Bitcoin mining involves several critical considerations:

Cost and ROI Analysis:

- **Initial Investment:** Assess the upfront costs of setting up the energy infrastructure, including hardware, installation, and any additional facilities.
- **Operational Costs:** Calculate ongoing costs, such as fuel, maintenance, and energy prices, to determine the total cost of ownership.
- **Return on Investment (ROI):** Perform a cost-benefit analysis to evaluate the potential returns from mining operations using different energy sources.

Environmental Impact:

- **Carbon Footprint:** Consider the greenhouse gas emissions associated with each energy source. Renewable energy sources generally have a lower environmental impact compared to fossil fuels.

- Sustainability: Evaluate the long-term sustainability of the energy source, including the availability of resources and potential environmental consequences.

Regulatory Compliance:

- **Legal Requirements:** Ensure that the chosen energy source complies with local, national, and international regulations and standards.
- **Incentives and Subsidies:** Investigate any government incentives or subsidies available for using renewable energy or implementing energy-efficient technologies.

Scalability and Long-Term Sustainability:

- **Capacity for Expansion:** Assess the potential for scaling up mining operations with the chosen energy source. Some energy sources, like renewable energy, may offer greater scalability.
- **Future-Proofing:** Consider the long-term viability of the energy source, including advancements in technology and potential changes in the regulatory landscape.

Demands for Powering Off Mining to Support the Local Grid in Times of Peak Energy Use:

- **Grid Support:** Some regions may require miners to power down during peak energy demand periods to support grid stability. Understand the local grid policies and their implications for mining operations.
- **Energy Flexibility:** Consider energy sources that can be easily managed or supplemented during peak demand periods, such as integrating battery storage systems with renewable energy.

The choice of energy source for Bitcoin mining significantly impacts both operational costs and environmental sustainability.

Grid electricity remains the most common energy source due to its availability and reliability, but its environmental impact varies based on the energy mix of the grid. Renewable energy sources, such as solar, wind, and hydro, offer sustainable alternatives that can reduce long-term costs and environmental impact but require significant initial investments and are subject to geographic and intermittent limitations.

Natural gas and nuclear power provide stable and high-efficiency energy solutions, though they come with environmental and safety challenges. Geothermal energy offers a sustainable option but is geographically limited and requires high upfront investment.

By understanding the different types of energy used in Bitcoin mining and their respective pros and cons, miners can make informed decisions that align with their operational goals and environmental responsibilities. Embracing innovative and sustainable energy solutions will not only enhance the efficiency and profitability of mining operations but also contribute to the broader goal of reducing the environmental impact of cryptocurrency mining.

Chapter 13:
Cooling Technologies

Effective cooling is critical in cryptocurrency mining to manage the heat generated by mining hardware. This chapter explores three primary cooling methods: Air Cooling, Immersion Cooling, and Hydro Cooling, each with its unique advantages and challenges.

Air Cooling involves using fans and heat sinks to dissipate heat. It is cost-effective, easy to implement, and widely available but can be noisy and less efficient in high-density setups.

Immersion Cooling submerges hardware in a dielectric liquid, offering high efficiency, reduced noise, and extended hardware lifespan. However, it has higher initial costs, requires specialized knowledge, and presents maintenance challenges.

Hydro Cooling uses water or water-based coolant with water blocks attached to components, providing excellent cooling efficiency and quieter operation. It supports overclocking but involves complex setup, maintenance, and the risk of leaks.

Understanding these cooling technologies helps miners optimize their operations, balance costs, and ensure their equipment's longevity and performance.

Let us look at each in more depth.

Air Cooling in Bitcoin Mining

Bitcoin mining has evolved into a sophisticated industry, where the efficiency and longevity of mining hardware are paramount. Among the various methods to manage the heat generated by mining rigs, air cooling remains the most traditional and widely used. This narrative delves into the intricacies of air cooling, exploring its mechanics, advantages, disadvantages, and practical applications in the dynamic world of bitcoin mining.

The Basics of Air Cooling

At its core, air cooling involves the use of fans and heat sinks to dissipate the heat produced by mining hardware. The primary goal is to maintain optimal temperatures to ensure the hardware operates efficiently and without thermal throttling, which can significantly reduce performance.

Fans are the workhorses of air-cooling systems. They function by moving air across heat sinks attached to the mining components, such as GPUs or ASICs. The heat sinks are designed with a large surface area to maximize heat transfer from the component to the air. As the air flows over the heat sinks, it carries the heat away, thereby cooling the components.

The simplicity of this system is one of its greatest strengths. The components needed for air cooling are readily available and easy to install, making it accessible to both novice and experienced miners.

Advantages of Air Cooling

Cost-Effectiveness: Air cooling systems are generally more affordable than other cooling solutions. The components, including fans and heat sinks, are inexpensive and widely available. This makes air cooling an attractive option for miners who are just starting out or those operating on a tight budget.

Ease of Implementation: Setting up an air-cooling system is straightforward. It requires minimal technical expertise, which is beneficial for miners who may not have advanced technical skills. The simplicity of air-cooling systems also means they are easier to maintain and troubleshoot compared to more complex cooling solutions.

Wide Availability of Components: The market is flooded with various fans and heat sinks, offering a range of options to suit different mining setups. Whether you need a high-performance fan for a densely packed mining rig or a quieter option for a home setup, you can easily find components that meet your requirements.

Flexibility and Scalability: Air cooling systems can be easily scaled to accommodate growing mining operations. As you add more mining rigs, you can simply install additional fans and heat sinks to manage the increased heat output. This flexibility makes air cooling a versatile solution for miners of all sizes.

Disadvantages of Air Cooling

Limited Efficiency in High-Density Setups: Air cooling has its limitations, particularly in high-density mining setups. As the number of mining rigs increases, the heat output can exceed the capacity of air-cooling systems to manage it effectively. This can lead to overheating, which reduces the performance and lifespan of the hardware.

Noise Levels: One of the main drawbacks of air cooling is the noise generated by the fans. In a typical mining setup, multiple fans are required to cool the hardware effectively. The combined noise from these fans can be significant, which can be disruptive, especially in home mining environments.

Higher Energy Consumption: Air cooling systems can be less energy-efficient compared to other cooling methods. Fans consume electricity, and in a large mining operation, this can add up to a substantial portion of the overall energy consumption. Higher energy use translates to increased operational costs, which can impact the profitability of mining operations.

Optimizing Air Cooling in Bitcoin Mining

To maximize the effectiveness of air-cooling systems, miners can implement several strategies:

Strategic Placement of Fans: The placement of fans within the mining rig and the surrounding environment is crucial for efficient cooling. Intake fans should be positioned to draw cool air into the system, while exhaust fans should be placed to expel hot air. This creates a continuous flow of air, ensuring that heat is effectively dissipated.

Regular Maintenance: Keeping fans and heat sinks clean is essential for maintaining their performance. Dust and debris can accumulate over time, reducing the efficiency of heat transfer. Regular cleaning and maintenance help to ensure that the cooling system operates at peak efficiency.

Use of High-Quality Components: Investing in high-quality fans and heat sinks can significantly improve the performance of air-cooling systems. High-performance fans can move more air at lower noise levels, and advanced heat sink designs can enhance heat dissipation.

Environmental Control: The ambient temperature and airflow in the mining environment play a significant role in the effectiveness of air cooling. Miners should aim to keep the mining area cool and well-ventilated. This can be achieved through the use of additional fans, air conditioning, or optimizing the layout of the mining rigs to enhance airflow.

Case Studies: Air Cooling in Action

Case Study 1: Home Mining Setup

John, a home miner, operates a small mining rig with three GPUs. He initially struggled with overheating issues, which impacted his mining performance. After researching cooling solutions, he decided to implement an air-cooling system.

John installed two high-performance intake fans at the front of his rig to draw in cool air and two exhaust fans at the rear to expel hot air. He also upgraded his heat sinks to models with larger surface areas for better heat dissipation. Additionally, John placed his mining rig in a well-ventilated area of his home and used a small portable air conditioner to keep the room temperature low.

The result was a significant reduction in hardware temperatures, leading to improved mining performance and stability. The noise from the fans was manageable, and the overall cost of the cooling solution was within his budget.

Case Study 2: Small-Scale Commercial Mining Operation

Sarah operates a small-scale commercial mining operation with 20 ASIC miners. Initially, she used a basic air-cooling setup with standard fans, but as she expanded her operation, she encountered overheating problems.

To address this, Sarah redesigned her cooling system. She installed industrial-grade fans capable of moving larger volumes of air and positioned them strategically to optimize airflow. Intake fans were placed on the sides of the mining rigs, while powerful exhaust fans were installed at the top to remove hot air. Sarah also implemented a regular maintenance schedule to clean the fans and heat sinks.

The enhanced air-cooling system allowed Sarah to maintain optimal temperatures across her mining operation. Although the noise levels increased, the improved cooling efficiency outweighed this drawback. The investment in higher-quality components and strategic planning paid off, resulting in higher mining efficiency and hardware longevity.

Case Study 3: Large-Scale Mining Farm

A large-scale mining farm faced significant challenges with heat management due to the high density of their mining rigs. The initial air-cooling setup was insufficient, leading to frequent overheating and reduced mining efficiency.

The farm's management decided to undertake a comprehensive overhaul of their cooling system. They consulted with experts to design a custom air-cooling solution tailored to their specific needs. This included the installation of large, high-capacity industrial fans and advanced heat sinks with enhanced thermal conductivity.

To further improve cooling, the farm was divided into several zones, each with its own dedicated cooling system. This zoning approach allowed for more precise control of airflow and temperature. The farm also invested in building modifications to enhance natural ventilation, such as installing vents and optimizing the layout of the mining rigs.

The result was a dramatic improvement in cooling efficiency. The new system maintained stable temperatures even during peak mining periods, leading to increased productivity and reduced downtime. The initial investment in the custom cooling solution was substantial, but the long-term benefits in terms of performance and hardware longevity justified the cost.

Future Trends in Air Cooling

As the bitcoin mining industry continues to evolve, so too will the technologies and strategies used for cooling. Future trends in air cooling are likely to focus on improving efficiency and reducing energy consumption. Some potential developments include:

Advanced Fan Technologies: Innovations in fan design and materials could lead to quieter and more efficient fans. These advancements might include the use of new aerodynamic designs, improved bearings, and more durable materials.

Smart Cooling Systems: Integration of smart technologies into cooling systems could allow for more precise control of temperatures. Smart fans equipped with sensors could automatically adjust their speed based on real-time temperature data, optimizing cooling efficiency and reducing energy consumption.

Enhanced Heat Sink Designs: Research into new materials and designs for heat sinks could yield components with superior thermal conductivity. This would improve the ability of heat sinks to dissipate heat, enhancing the overall performance of air-cooling systems.

Hybrid Cooling Solutions: Combining air cooling with other cooling methods, such as liquid cooling, could provide a hybrid solution that leverages the strengths of both systems. This approach could offer higher efficiency and flexibility, particularly in large-scale mining operations.

Air cooling remains a cornerstone of bitcoin mining due to its cost-effectiveness, ease of implementation, and flexibility. While it has its limitations, particularly in high-density setups, strategic planning and the use of high-quality components can significantly enhance its performance. As the mining industry progresses, ongoing innovations in cooling technologies will continue to improve the efficiency and sustainability of mining operations. By understanding and optimizing air cooling systems, miners can ensure their hardware operates at peak efficiency, maximizing their mining potential and profitability.

Immersion Cooling in Bitcoin Mining: A Comprehensive Guide

As bitcoin mining has evolved into a highly competitive industry, miners are continually seeking ways to optimize their operations and maximize efficiency. One of the most advanced and effective cooling methods that has gained popularity in recent years is immersion cooling. This narrative explores the fundamentals of immersion cooling, its advantages and disadvantages, and its practical applications in the field of bitcoin mining.

The Basics of Immersion Cooling

Immersion cooling involves submerging mining hardware in a dielectric (non-conductive) liquid. Unlike traditional air cooling, which relies on fans and heat sinks to dissipate heat, immersion cooling allows the liquid to directly absorb the heat generated by the hardware. The heated liquid is then circulated through a cooling system, where the heat is dissipated, and the cooled liquid is returned to the hardware.

There are two main types of immersion cooling: single-phase and two-phase. In single-phase immersion cooling, the liquid remains in a liquid state throughout the cooling process. In two-phase immersion cooling, the liquid changes from a liquid to a gas as it absorbs heat and then condenses back into a liquid in the cooling system.

Advantages of Immersion Cooling

High Cooling Efficiency: Immersion cooling is highly efficient at transferring heat away from mining hardware. The dielectric liquid used in the process has a much higher heat capacity and thermal conductivity than air, allowing for more effective cooling. This efficiency enables miners to maintain optimal temperatures even in densely packed mining setups.

Reduced Noise: Since immersion cooling does not rely on fans, it significantly reduces the noise levels associated with traditional air-cooling systems. This makes immersion cooling an attractive option for miners who need a quieter operating environment.

Extended Hardware Lifespan: Maintaining optimal temperatures with immersion cooling can extend the lifespan of mining hardware. By preventing overheating and reducing thermal stress, immersion cooling helps to preserve the integrity and performance of the components over time.

Higher Density Mining: The superior cooling capabilities of immersion cooling allow for higher density mining setups. Miners can pack more hardware into a smaller space without worrying about overheating, maximizing their use of available space and increasing their overall mining capacity.

Energy Efficiency: Immersion cooling can be more energy-efficient than air cooling. The absence of fans reduces energy consumption, and the efficient heat transfer properties of the dielectric liquid further contribute to lower energy use. This can result in significant cost savings for large-scale mining operations.

Disadvantages of Immersion Cooling

High Initial Setup Cost: The initial cost of setting up an immersion cooling system can be substantial. The specialized equipment, including tanks, pumps, and cooling systems, as well as the dielectric liquid itself, can be expensive. This higher upfront investment may be a barrier for smaller mining operations.

Specialized Equipment and Knowledge: Implementing and maintaining an immersion cooling system requires specialized equipment and expertise. Miners need to have a thorough understanding of the cooling process and the materials involved to ensure the system operates effectively and safely.

Maintenance and Handling Challenges: The maintenance of immersion cooling systems can be more complex than traditional air cooling. Handling the dielectric liquid, ensuring the system remains leak-free, and performing regular maintenance checks require diligence and expertise. Additionally, any hardware upgrades or replacements necessitate careful handling of the liquid and components.

Optimizing Immersion Cooling in Bitcoin Mining

To maximize the benefits of immersion cooling, miners can implement several strategies:

Choosing the Right Dielectric Liquid: Selecting the appropriate dielectric liquid is crucial for the effectiveness of the cooling system. The liquid should have high thermal conductivity, be chemically stable, and be compatible with the mining hardware. Commonly used dielectric liquids include mineral oils, synthetic fluids, and specialized cooling fluids designed for immersion cooling.

System Design and Layout: Designing the immersion cooling system with careful consideration of the layout and flow of the dielectric liquid is essential. Ensuring proper circulation and heat dissipation can optimize the cooling performance. Using multiple tanks or chambers to separate different stages of the cooling process can enhance efficiency.

Regular Maintenance and Monitoring: Regular maintenance and monitoring are vital to keep the immersion cooling system running smoothly. This includes checking for leaks, monitoring the temperature and flow of the dielectric liquid, and performing periodic cleaning of the system components.

Environmental Control: The ambient temperature and airflow in the mining environment can impact the performance of the immersion cooling system. Maintaining a cool and well-ventilated environment can help to enhance the overall cooling efficiency.

Case Studies: Immersion Cooling in Action

Case Study 1: Small-Scale Mining Setup

David, a home miner, decided to transition to immersion cooling to address the overheating issues he faced with his traditional air-cooling setup. He set up a single-phase immersion cooling system using mineral oil and a custom-built cooling tank.

David carefully selected high-quality mineral oil with excellent thermal properties and designed his tank to ensure optimal circulation of the liquid. He installed a pump to circulate the heated oil through a radiator system, where it was cooled before returning to the tank.

The result was a significant reduction in hardware temperatures and a quieter operating environment. David was able to increase his mining efficiency and extend the lifespan of his hardware, making the initial investment in immersion cooling worthwhile.

Case Study 2: Medium-Scale Commercial Mining Operation

Emma runs a medium-scale commercial mining operation with 50 ASIC miners. As her operation grew, she encountered significant heat management challenges with her air-cooling system. After researching alternative solutions, she decided to implement a two-phase immersion cooling system.

Emma collaborated with a specialized cooling solutions provider to design and install the system. The setup involved submerging the ASIC miners in a specialized dielectric fluid that vaporizes as it absorbs heat. The vapor was then condensed back into a liquid in a cooling tower and recirculated to the tanks.

The two-phase immersion cooling system provided superior cooling efficiency, allowing Emma to maintain stable temperatures across her mining operation. Despite the higher initial costs and complexity of the system, the long-term benefits of increased mining efficiency and hardware longevity justified the investment.

Case Study 3: Large-Scale Mining Farm

A large-scale mining farm faced severe overheating issues due to the high density of their mining rigs. The management decided to overhaul their cooling system and opted for a large-scale immersion cooling solution.

The farm worked with experts to design a custom immersion cooling system that included multiple tanks and advanced circulation systems. They used high-performance synthetic fluids specifically designed for immersion cooling, which provided excellent thermal conductivity and chemical stability.

The new system allowed the farm to significantly increase its mining capacity by accommodating more rigs in a smaller space. The reduced energy consumption and lower maintenance costs further improved the farm's profitability. The investment in the immersion cooling system paid off, resulting in higher mining efficiency and reduced downtime.

Future Trends in Immersion Cooling

As the bitcoin mining industry continues to evolve, several trends are likely to shape the future of immersion cooling:

Development of Advanced Cooling Fluids: Ongoing research and development are expected to yield new and improved cooling fluids with higher thermal conductivity and better chemical stability. These advanced fluids will enhance the performance and efficiency of immersion cooling systems.

Integration of Smart Technologies: The integration of smart technologies, such as IoT sensors and automated control systems, will allow for more precise monitoring and management of immersion cooling systems. These technologies can optimize cooling performance, reduce energy consumption, and simplify maintenance tasks.

Hybrid Cooling Solutions: Hybrid cooling solutions that combine immersion cooling with other cooling methods, such as air or liquid cooling, may offer enhanced flexibility and efficiency. These systems can leverage the strengths of different cooling techniques to achieve optimal performance.

Increased Adoption in Large-Scale Operations: As immersion cooling technology becomes more accessible and cost-effective, its adoption in large-scale mining operations is expected to increase.

The superior cooling efficiency and potential cost savings make it an attractive option for mining setups.

Immersion cooling represents a cutting-edge solution for managing the heat generated by bitcoin mining hardware. Its high cooling efficiency, reduced noise levels, and ability to support high-density mining setups make it an appealing choice for miners seeking to optimize their operations. While the initial setup costs and maintenance challenges can be significant, the long-term benefits of improved mining efficiency and hardware longevity often outweigh these drawbacks.

As the industry continues to innovate, immersion cooling is poised to play a crucial role in the future of bitcoin mining. By understanding and implementing effective immersion cooling strategies, miners can enhance their operational efficiency, reduce energy consumption, and maximize the profitability of their mining endeavors.

Hydro Cooling in Bitcoin Mining

Bitcoin mining is an energy-intensive process that generates significant heat. Efficient cooling systems are essential to maintain optimal performance and prolong the lifespan of mining hardware. One of the advanced methods of cooling that has gained traction in the mining community is hydro cooling, also known as water cooling. This guide explores the fundamentals of hydro cooling, its advantages and disadvantages, and its practical applications in bitcoin mining.

The Basics of Hydro Cooling

Hydro cooling involves using water or a water-based coolant to transfer heat away from mining hardware. This method typically employs water blocks attached to the components that generate the most heat, such as GPUs or ASICs. These water blocks absorb heat from the hardware and transfer it to the coolant. The heated coolant is then pumped through a radiator system, where the heat is dissipated into the surrounding air before the cooled liquid returns to the water blocks.

The efficiency of hydro cooling lies in water's high heat capacity and thermal conductivity compared to air. This allows for more effective heat transfer and cooling performance, making hydro cooling a viable option for high-density mining operations.

Advantages of Hydro Cooling

Excellent Cooling Efficiency: Hydro cooling is highly efficient due to water's superior thermal conductivity and heat capacity. This method can maintain lower temperatures for mining hardware, preventing thermal throttling and enhancing performance.

Quieter Operation: Compared to air cooling systems, which rely on multiple fans, hydro cooling systems are generally quieter. The use of pumps and radiators reduces the overall noise level, making it a more suitable option for miners who need a quieter working environment.

Support for Overclocking: The enhanced cooling provided by hydro cooling allows miners to safely overclock their hardware, potentially increasing mining performance. By maintaining lower temperatures, hydro cooling reduces the risk of overheating and hardware damage associated with overclocking.

Scalability: Hydro cooling systems can be scaled to accommodate larger mining operations. By adding more water blocks, radiators, and pumps, miners can expand their cooling system to match the growing demands of their mining setups.

Energy Efficiency: Although hydro cooling systems require pumps to circulate the coolant, they can be more energy-efficient than air cooling systems in high-density setups. The efficient heat transfer properties of water reduce the overall energy consumption needed to maintain optimal temperatures.

Disadvantages of Hydro Cooling

Complexity of Setup and Maintenance: Installing and maintaining a hydro cooling system is more complex than air cooling. The system requires careful planning and assembly, including securing water blocks, connecting tubing, and setting up pumps and radiators. Regular maintenance, such as checking for leaks and cleaning components, is also necessary to ensure optimal performance.

Risk of Leaks: One of the significant risks associated with hydro cooling is the potential for leaks. Water leaks can cause severe damage to mining hardware and electrical components. Ensuring leak-proof connections and regularly inspecting the system are essential to mitigate this risk.

Higher Initial Costs: Hydro cooling systems have higher initial setup costs compared to air cooling. The cost of water blocks, radiators, pumps, and coolant can add up, making it a more substantial investment. However, the long-term benefits of improved cooling efficiency and hardware longevity may offset these initial costs.

Space Requirements: Hydro cooling systems require additional space for radiators, pumps, and tubing. This can be a limitation for miners with constrained physical space. Proper planning and layout optimization are necessary to accommodate the cooling system without hindering the overall mining operation.

Optimizing Hydro Cooling in Bitcoin Mining

To maximize the benefits of hydro cooling, miners can implement several strategies:

Selecting High-Quality Components: Investing in high-quality water blocks, pumps, and radiators is crucial for the performance and reliability of the hydro cooling system. Components with better thermal conductivity and durability can enhance cooling efficiency and reduce the risk of leaks.

Proper System Design: Designing the hydro cooling system with optimal flow paths and component placement is essential. Ensuring that the coolant flows efficiently through the system and effectively dissipates heat from the hardware can maximize cooling performance.

Regular Maintenance: Regular maintenance is critical to the longevity and performance of hydro cooling systems. This includes checking for leaks, cleaning radiators and water blocks, and replacing the coolant periodically. A well-maintained system can operate efficiently and reduce the risk of hardware damage.

Monitoring and Control: Implementing monitoring systems to track temperatures, flow rates, and coolant levels can help miners maintain optimal operating conditions. Automated controls can adjust pump speeds and fan settings based on real-time temperature data, enhancing cooling efficiency.

Environmental Considerations: The ambient temperature and ventilation in the mining environment impact the performance of hydro cooling systems. Keeping the mining area cool and well-ventilated can enhance the efficiency of the radiators and overall cooling system.

Case Studies: Hydro Cooling in Action

Case Study 1: Home Mining Setup

Mike, a home miner, decided to upgrade his cooling system to hydro cooling to address the noise and inefficiency of his air-cooled setup. He installed a hydro cooling system with water blocks on his GPUs, a pump, and a radiator mounted outside his mining rig.

The hydro cooling system significantly reduced the temperatures of Mike's GPUs, allowing him to overclock his hardware safely. The noise levels also dropped, creating a more comfortable environment for home mining. Although the initial setup required careful planning and investment, the improved performance and quieter operation justified the switch to hydro cooling.

Case Study 2: Small-Scale Commercial Mining Operation

Laura runs a small-scale commercial mining operation with 30 ASIC miners. She faced challenges with heat management and noise using air cooling systems. After researching alternatives, Laura opted for hydro cooling to improve efficiency and reduce noise.

Working with a cooling solutions provider, Laura installed a hydro cooling system with custom water blocks for her ASIC miners, multiple radiators, and high-flow pumps. The system was designed to ensure efficient coolant circulation and heat dissipation.

The hydro cooling system provided superior cooling, maintaining stable temperatures across Laura's mining operation. The reduced noise levels and increased energy efficiency enhanced the working environment and operational profitability. Despite the higher initial costs, the long-term benefits of hydro cooling were evident in improved mining performance and hardware longevity.

Case Study 3: Large-Scale Mining Farm

A large-scale mining farm faced significant heat management challenges due to the high density of their mining rigs. The management decided to implement a comprehensive hydro cooling solution to address these issues.

The farm collaborated with experts to design and install a custom hydro cooling system. The setup included high-performance water blocks, industrial-grade pumps, and large radiators to handle the extensive heat output. The system was segmented into zones, each with dedicated cooling loops to optimize efficiency.

The hydro cooling system allowed the farm to increase its mining capacity by supporting more rigs in a smaller space. Efficient heat management reduced energy consumption and operational costs, while quieter operation improved the working environment. The initial investment in hydro cooling was substantial, but the long-term gains in performance and efficiency justified the expenditure.

Future Trends in Hydro Cooling

Several trends are likely to shape the future of hydro cooling in bitcoin mining:

Advanced Materials and Designs: Ongoing research into new materials and designs for water blocks and radiators is expected to yield components with superior thermal conductivity and durability. These advancements will enhance the performance and efficiency of hydro cooling systems.

Integration of Smart Technologies: The incorporation of smart technologies, such as IoT sensors and automated control systems, will enable more precise monitoring and management of hydro cooling systems. These technologies can optimize cooling performance, reduce energy consumption, and simplify maintenance tasks.

Hybrid Cooling Solutions: Hybrid cooling solutions that combine hydro cooling with other cooling methods, such as air or immersion cooling, may offer enhanced flexibility and efficiency. These systems can leverage the strengths of different cooling techniques to achieve optimal performance.

Increased Adoption in High-Density Operations: As hydro cooling technology becomes more accessible and cost-effective, its adoption in high-density mining operations is expected to increase. The superior cooling efficiency and potential cost savings make it an attractive option for large-scale mining setups.

Hydro cooling represents an advanced and efficient solution for managing the heat generated by bitcoin mining hardware. Its high cooling efficiency, quieter operation, and support for overclocking make it an appealing choice for miners seeking to optimize their operations. While the complexity of setup, risk of leaks, and higher initial costs present challenges, the long-term benefits of improved performance and hardware longevity often outweigh these drawbacks.

As the bitcoin mining industry continues to evolve, hydro cooling is poised to play a crucial role in the future of mining operations. By understanding and implementing effective hydro cooling strategies, miners can enhance their operational efficiency, reduce energy consumption, and maximize the profitability of their mining endeavors.

Chapter 14:
Safety and Compliance

Operational Safety Standards

Ensuring operational safety in Bitcoin mining is critical to protect both personnel and equipment. The high energy consumption and significant heat generation associated with mining operations pose various safety risks. Implementing comprehensive safety standards helps mitigate these risks and ensures smooth, uninterrupted mining operations.

Electrical Safety:

- **Proper Wiring:** Ensure that all electrical installations are performed by qualified professionals, using appropriate wiring standards to handle the high power load of mining equipment.
- **Circuit Protection:** Install circuit breakers and surge protectors to prevent electrical overloads and protect equipment from power surges.
- **Grounding:** Properly ground all electrical components to prevent electrical shocks and fires.

Fire Safety:

- **Fire Suppression Systems:** Install fire suppression systems, such as sprinklers or gas-based systems, to quickly extinguish fires in mining facilities.
- **Fire Extinguishers:** Place fire extinguishers in accessible locations throughout the mining facility and ensure staff are trained in their use.

- **Heat Management:** Implement effective cooling solutions to manage the heat generated by mining hardware and prevent overheating.

Physical Safety:

- **Equipment Placement:** Ensure mining equipment is placed in well-ventilated areas with adequate spacing to allow for airflow and heat dissipation.
- **Secure Mounting:** Securely mount all hardware to prevent accidents caused by equipment falling or shifting.
- **Personal Protective Equipment (PPE):** Provide PPE, such as gloves and safety glasses, to staff handling hardware and performing maintenance tasks.

Noise Management:

- **Soundproofing:** Implement soundproofing measures to reduce the noise levels generated by mining rigs and cooling systems.
- **Hearing Protection:** Provide hearing protection to personnel working in areas with high noise levels.

Electromagnetic Interference (EMI)

Bitcoin mining operations, especially large-scale ones, involve significant amounts of hardware, including powerful computers and specialized mining rigs. These operations generate substantial amounts of electricity, which in turn creates electromagnetic fields (EMF).

The risks of magnetic issues surrounding Bitcoin miners can be understood in several ways:

- **Risk:** The high concentration of electronic equipment in Bitcoin mining farms can generate substantial EMI, which can interfere with nearby electronic devices, communications systems, and potentially even sensitive industrial or medical equipment.

- **Mitigation**: Shielding, proper grounding, and adhering to EMI regulations can minimize these risks.

Health Concerns

- **Risk**: Prolonged exposure to strong electromagnetic fields can potentially pose health risks to individuals working in close proximity to mining equipment. There is ongoing debate about the long-term effects of EMF exposure, though most regulatory bodies deem the risks minimal if exposure is within recommended limits.
- **Mitigation**: Implementing EMF shielding and ensuring that employees follow safety guidelines can reduce these potential risks.

Magnetic Field Disruption

- **Risk**: Magnetic fields generated by mining operations might interfere with nearby magnetic storage devices or other magnetically sensitive equipment. This could lead to data loss or equipment malfunction.
- **Mitigation**: Keeping sensitive equipment at a safe distance from mining operations and using non-magnetic materials in construction can help mitigate these issues.

Environmental Impact

- **Risk**: The energy consumption of Bitcoin miners contributes to environmental degradation, and the generation of strong EMFs is a byproduct of this high energy use. The broader environmental implications of such operations include increased energy demand, which can exacerbate issues like climate change if the energy is sourced from fossil fuels.

- **Mitigation**: Using renewable energy sources and implementing energy-efficient practices can reduce the environmental footprint of Bitcoin mining.

Regulatory and Compliance Risks

- **Risk**: As awareness of EMF and its potential impacts grows, there could be increased regulatory scrutiny on Bitcoin mining operations, particularly concerning their EMF emissions. Failure to comply with regulations could result in fines or forced shutdowns.
- **Mitigation**: Staying informed about local and international EMF regulations and ensuring compliance can help avoid regulatory risks.

Summary Table

Risk	Description	Mitigation
Electromagnetic Interference	EMI can disrupt nearby electronic devices.	Shielding, grounding, and adherence to EMI standards
Health Concerns	Prolonged exposure to strong EMFs may pose health risks.	EMF shielding, safety guidelines adherence
Magnetic Field Disruption	Magnetic fields might interfere with sensitive equipment.	Distance sensitive equipment, use non-magnetic materials

Environmental Impact	High energy use contributes to environmental degradation.	Use renewable energy, improve energy efficiency
Regulatory and Compliance Risks	Potential increased regulatory scrutiny on EMF emissions.	Ensure compliance with EMF regulations

Understanding and managing these risks is crucial for the safe and efficient operation of Bitcoin mining activities, especially as the industry continues to expand.

EMF vs EMI

Electromagnetic Fields (EMF) and Electromagnetic Interference (EMI) are related but distinct concepts. EMF refers to the physical fields produced by electrically charged objects, encompassing both electric and magnetic fields, which are naturally occurring or generated by electronic devices.

EMI, on the other hand, is a disturbance caused by an external electromagnetic field that disrupts the normal operation of an electronic device. While EMF is the broader phenomenon, EMI is a specific effect where unwanted EMF interferes with the functioning of other electronic equipment, potentially causing malfunctions or data loss.

While we are on this topic, we thought it would be entertaining to hear a personal story.

In the very early days of our Bitcoin mining operation, we were full of excitement and ambition, diving headfirst into the world of cryptocurrency with little understanding of the potential pitfalls that lay ahead. The rows of powerful computers and specialized mining rigs buzzed with activity, solving complex cryptographic puzzles around the clock. But as the machines hummed away, we quickly discovered that this intense operation came with unforeseen challenges.

Our first encounter with Electromagnetic Interference (EMI) was a harsh lesson. We had set up our monitors right next to the mining rigs, blissfully unaware of the invisible forces at play. It wasn't long before we started experiencing strange flickers and then, suddenly, both monitors went dark—completely fried by the powerful EMI generated by our equipment. Our first thoughts were that the monitors were defective. We replaced them to only encounter the same issue. Now 4 monitors were lost.

The loss was a wake-up call. We realized that our setup was far from ideal, and we needed to rethink our approach if we were going to avoid further damage to our equipment..

The solution wasn't straightforward, but it was necessary. We ended up creating a shielded cable to protect our equipment from the overwhelming EMI and moved the monitors to a different room, far from the mining rigs. It was a simple but effective fix, and it marked the beginning of our education in the electromagnetic complexities of large-scale mining.

But the EMI didn't just wreak havoc on our monitors—it even took out our wallets, in a manner of speaking. One day, after spending hours near the mining rigs, we found ourselves at a local store unable to make a single purchase. All of our credit cards had become useless, their magnetic strips scrambled by the same invisible fields that had destroyed our monitors. That was a particularly frustrating lesson, but it drove home the importance of respecting the power of the electromagnetic fields surrounding us.

From that point on, we made it a rule: wallets and phones were to be left outside the mining room. It became part of our routine—no one entered the rig room with anything that had a magnetic strip or electronic component. It was a small inconvenience compared to the damage we had experienced, and it ensured that we didn't repeat the same mistakes.

These early challenges taught us more than we ever expected about the risks of magnetic fields in Bitcoin mining. We realized that this wasn't just about protecting our equipment; it was about understanding the environment we were creating and taking the necessary steps to mitigate any potential harm. Whether it was constructing shielded cables, relocating sensitive equipment, or simply leaving our wallets and phones outside, each step was a move toward a safer and more reliable operation.

Looking back, these experiences were pivotal. They not only saved us from further losses but also instilled in us a deep respect for the complexities of running a mining operation. As our understanding grew, so did our ability to manage the myriad risks that came with the territory, ensuring that our journey into the world of Bitcoin mining was as safe and sustainable as it was innovative.

Regulatory Compliance

Compliance with local, national, and international regulations is essential for the legal operation of Bitcoin mining activities. Regulatory requirements can vary significantly across different jurisdictions, and miners must stay informed and adhere to applicable laws and standards.

Permits and Licenses:

- **Business Licensing:** Obtain the necessary business licenses to operate a mining facility legally.
- **Building Permits:** Ensure that any construction or modification of mining facilities complies with local building codes and regulations.

- **Environmental Permits:** Acquire permits for environmental impact assessments and waste management, if required.

Energy Regulations:

- **Electricity Use:** Adhere to regulations governing electricity consumption, including peak demand restrictions and energy efficiency standards.
- Renewable Energy Incentives: Take advantage of incentives and subsidies for using renewable energy sources, where available.

Tax Compliance:

- **Reporting Earnings:** Accurately report mining earnings and adhere to tax regulations for cryptocurrency transactions.
- **Tax Deductions:** Keep detailed records of expenses related to mining operations to claim eligible tax deductions.

Anti-Money Laundering (AML) and Know Your Customer (KYC):

- **AML Compliance:** Implement AML procedures to prevent the use of mining operations for money laundering activities.
- **KYC Procedures:** If providing mining services to others, implement KYC procedures to verify the identity of clients and comply with financial regulations.

Environmental Considerations

The environmental impact of Bitcoin mining is a growing concern, given the high energy consumption and carbon footprint associated with mining activities. Implementing environmentally responsible practices helps minimize the ecological impact and contributes to sustainability.

Energy Efficiency:

- **Efficient Hardware:** Use energy-efficient mining hardware to reduce electricity consumption and operating costs.
- **Optimized Cooling:** Implement advanced cooling solutions, such as liquid cooling or immersion cooling, to improve energy efficiency and reduce heat output.

Renewable Energy:

- **Solar and Wind Power:** Invest in solar panels or wind turbines to generate renewable energy for mining operations, reducing reliance on fossil fuels.
- **Hydropower and Geothermal Energy:** Leverage hydroelectric or geothermal power sources, where available, to minimize environmental impact.

Carbon Offsetting:

- **Carbon Credits:** Purchase carbon credits to offset the carbon emissions generated by mining operations.
- **Reforestation Projects:** Invest in reforestation projects or other environmental initiatives to balance the ecological footprint of mining activities.

Waste Management:

- E-Waste Recycling: Implement e-waste recycling programs to responsibly dispose of and recycle outdated or damaged mining hardware.
- **Hazardous Materials:** Properly handle and dispose of hazardous materials, such as batteries and electronic components, in compliance with environmental regulations.

Community Engagement:

- **Local Collaboration:** Engage with local communities to address environmental concerns and collaborate on sustainability initiatives.
- **Public Awareness:** Raise awareness about the environmental impact of Bitcoin mining and promote responsible mining practices.

Ensuring safety and compliance in Bitcoin mining operations is critical to protect personnel, equipment, and the environment. Implementing comprehensive safety standards, adhering to regulatory requirements, and adopting environmentally responsible practices contribute to the long-term sustainability and legality of mining activities.

Operational safety standards, such as electrical and fire safety measures, physical safety protocols, and noise management, help mitigate risks and protect mining facilities. Regulatory compliance with permits, licenses, energy regulations, tax laws, and AML/KYC procedures ensures the legal operation of mining activities.

Environmental considerations, including energy efficiency, renewable energy adoption, carbon offsetting, waste management, and community engagement, help minimize the ecological impact of mining operations. By prioritizing safety, compliance, and sustainability, miners can achieve successful and responsible Bitcoin mining operations.

Part III:
Advanced Mining Strategies and Considerations

Chapter 15:
Mining Hash Rate Optimization

Optimizing the hash rate of your Bitcoin mining operations is paramount to maximizing profitability and ensuring long-term sustainability. Hash rate, a measure of the computational power dedicated to mining, directly influences the probability of solving cryptographic puzzles and earning mining rewards. This chapter delves into the importance of hash rate, techniques to optimize it, the intricacies of overclocking and undervolting, and the continuous efforts required to monitor and maintain optimal performance.

Importance of Hash Rate in Mining

The hash rate is a critical metric in the realm of cryptocurrency mining. It represents the number of calculations your mining hardware can perform per second. A higher hash rate improves your chances of successfully adding a block to the blockchain and receiving the associated rewards.

Mining Rewards: The primary incentive for increasing your hash rate is to secure more mining rewards. Bitcoin miners are rewarded with newly minted coins and transaction fees for validating blocks. As the hash rate increases, so does the probability of earning these rewards, making your mining operation more profitable.

Network Security: The aggregate hash rate of all miners in the network enhances the security of the blockchain. A higher overall hash rate makes the network more resilient to attacks, such as the 51% attack, where a single entity could potentially control the majority of the network's computational power. By contributing to a higher hash rate, miners play a crucial role in maintaining the integrity and security of the blockchain.

Profitability: In a competitive mining environment, optimizing hash rate is essential for profitability. As mining difficulty increases with the growing network hash rate, miners must continually improve their hardware and software to stay profitable. Efficiently managing and maximizing your hash rate helps ensure a favorable return on investment (ROI).

Techniques to Optimize Your Mining Hash Rate

Optimizing hash rate involves a combination of hardware and software adjustments, strategic decisions, and continuous monitoring. Here are some effective techniques to enhance your mining hash rate:

Choose the Right Hardware:

- **ASICs vs. GPUs:** Application-Specific Integrated Circuits (ASICs) are designed specifically for mining and offer higher hash rates and energy efficiency compared to Graphics Processing Units (GPUs). Selecting the appropriate hardware based on the cryptocurrency being mined and budget constraints is crucial.
- **Latest Models:** Investing in the latest models of mining hardware ensures better performance and energy efficiency. Manufacturers continually release improved versions of ASICs with higher hash rates and lower power consumption.

Optimize Mining Software:

- **Software Selection:** Using mining software optimized for your specific hardware and the cryptocurrency you are mining can significantly impact performance. Popular mining software includes CGMiner, BFGMiner, and NiceHash, each offering different features and optimization capabilities.
- **Algorithm Tuning:** Adjusting software settings to optimize performance for the specific mining algorithm (e.g., SHA-

144

256 for Bitcoin, Ethash for Ethereum Classic) can boost efficiency. This includes fine-tuning parameters such as intensity, thread concurrency, and clock speeds.

Optimize Network and Infrastructure:

- **Stable Internet Connection:** Ensuring a stable and high-speed internet connection minimizes downtime and latency, which are critical for maintaining a high hash rate. Wired connections (Ethernet) are preferable over wireless to reduce network interruptions.
- **Power Supply:** Using high-quality, efficient power supplies ensures consistent and stable power delivery to your mining hardware, preventing performance drops and hardware damage.

Join a Mining Pool:

- **Pool Selection:** Joining a mining pool allows you to combine your hash rate with other miners, increasing the chances of earning rewards. Selecting a reputable pool with low fees, reliable payouts, and a good track record is essential for maximizing profitability.
- **Pool Configuration:** Properly configuring your mining software to connect to the pool ensures optimal performance and contribution to the pool's overall hash rate.

Overclocking and Undervolting

Overclocking and undervolting are advanced techniques used to optimize the performance and efficiency of mining hardware. These methods require a careful balance to maximize hash rate while maintaining stability and minimizing power consumption.

Overclocking:

Overclocking involves increasing the clock speed of the GPU or ASIC beyond the manufacturer's recommended settings to achieve higher hash rates.

Benefits:

- **Increased Performance:** Higher clock speeds can significantly boost the hash rate, leading to increased mining rewards.
- **Competitive Edge:** Overclocking can provide a competitive edge by maximizing the performance of your existing hardware without immediate additional investment.

Risks:

- **Heat Generation:** Overclocking generates more heat, which can reduce hardware lifespan if not managed properly.
- **Stability Issues:** Pushing hardware beyond its intended limits can lead to instability and crashes.
- **Warranty Void:** Overclocking often voids the manufacturer's warranty, leaving you responsible for any damage.

Steps to Overclock:

1. **Benchmarking:** Start by benchmarking your hardware to determine its current performance and stability.
2. **Incremental Adjustments:** Gradually increase the clock speed in small increments, testing stability and performance after each adjustment.
3. **Cooling Solutions:** Ensure adequate cooling to manage the additional heat generated by overclocking. This might involve upgrading existing cooling systems or implementing advanced solutions like liquid cooling.

4. **Monitoring:** Continuously monitor hardware temperatures and stability to prevent overheating and hardware damage.

Undervolting:

Undervolting involves reducing the voltage supplied to the GPU or ASIC while maintaining stable performance. This technique aims to improve energy efficiency and reduce heat output.

Benefits:

- **Lower Power Consumption:** Reducing voltage lowers power consumption, decreasing electricity costs and improving energy efficiency.
- **Reduced Heat Generation:** Less voltage means less heat, which can extend hardware lifespan and reduce cooling requirements.
- **Stability:** Proper undervolting can maintain or even improve stability by preventing thermal throttling.

Risks:

- **Instability:** Insufficient voltage can lead to instability and crashes, affecting mining efficiency.
- **Complexity:** Finding the optimal undervolt setting requires time and experimentation.

Steps to Undervolt:

1. **Software Tools:** Use software tools like MSI Afterburner (for GPUs) to adjust voltage settings.
2. **Gradual Reduction:** Gradually decrease the voltage in small increments while testing stability and performance.
3. **Monitoring:** Monitor hardware temperatures and performance metrics to ensure stable operation at reduced voltage levels.

Monitoring and Maintaining Optimal Performance

Continuous monitoring and maintenance are essential to sustain optimal performance and longevity of your mining hardware. Implementing the following practices helps maintain peak performance and prevent costly downtimes.

Real-Time Monitoring:

- **Monitoring Tools:** Use comprehensive monitoring software to track key performance metrics such as hash rate, temperature, power consumption, and uptime. Popular tools include Hive OS, Awesome Miner, and Minerstat.
- **Alerts and Notifications:** Set up alerts and notifications for critical events, such as hardware failures, temperature spikes, and network issues. Prompt notifications enable quick responses to potential problems.

Regular Maintenance:

- **Cleaning:** Regularly clean mining hardware to prevent dust buildup, which can impede cooling and reduce efficiency. Dust accumulation can cause overheating and hardware failures.
- **Thermal Paste:** Reapply thermal paste on GPUs or ASICs as needed to ensure effective heat dissipation. Over time, thermal paste can dry out, reducing its effectiveness.
- **Firmware Updates:** Keep mining hardware and software up to date with the latest firmware and software updates to benefit from performance improvements and security patches. Manufacturers often release updates to fix bugs, enhance stability, and improve efficiency.

Performance Optimization:

- **Load Balancing:** Distribute the workload evenly across all mining rigs to prevent overloading and ensure consistent performance. This helps in optimizing the use of available resources.

- **Cooling Solutions:** Optimize cooling solutions to maintain safe operating temperatures and prevent thermal throttling. Efficient cooling enhances performance and prolongs hardware lifespan.

Energy Management:

- **Energy Efficiency:** Continuously seek ways to improve energy efficiency, such as optimizing power settings and using energy-efficient hardware. Lower energy consumption translates to reduced operational costs.
- **Power Management:** Monitor power usage and adjust settings to balance performance and energy consumption. Implementing power-saving measures during low network difficulty periods can save on electricity costs.

Optimizing hash rate is a critical aspect of maximizing profitability in Bitcoin mining. By selecting the right hardware, optimizing mining software, and employing techniques like overclocking and undervolting, miners can enhance their hash rate and improve their mining efficiency. Continuous monitoring and regular maintenance are essential to sustain optimal performance and ensure the longevity of mining hardware.

As the cryptocurrency mining landscape evolves, staying informed about the latest advancements and best practices in hash rate optimization will be key to maintaining a competitive edge and achieving long-term success in mining operations. By adopting a proactive approach to optimization and maintenance, miners can navigate the challenges of a dynamic industry and secure their place in the future of cryptocurrency mining.

Chapter 16:
Scaling Your Mining Operations

Scaling up your Bitcoin mining operations is not merely a matter of adding more rigs. It involves strategic planning, investment in infrastructure, optimization of power and cooling, and meticulous management of increased maintenance and monitoring tasks. This chapter explores these aspects in detail, providing a comprehensive guide to successfully expanding your mining activities.

Strategies for Scaling Up Your Mining Activities

Assessing and Planning:

Scaling your mining operation begins with a thorough assessment and meticulous planning. A feasibility study is essential to evaluate the economic viability of expansion. This involves analyzing current market conditions, electricity costs, hardware prices, and potential profitability. A feasibility study should also consider the long-term sustainability of the project, potential risks, and the regulatory environment. Understanding these factors helps in making informed decisions and avoiding costly mistakes.

Once the feasibility study is complete, develop a comprehensive scalability plan. This plan should outline the steps and resources required for expansion, including timelines, budget allocations, and key performance indicators (KPIs). The plan should also address potential challenges and strategies to overcome them. By having a clear roadmap, you can systematically approach scaling your operations without being overwhelmed by the complexities involved.

Incremental Expansion:

A phased approach to scaling is often the most effective strategy. Incremental expansion allows you to gradually increase the number of mining rigs and infrastructure, making it easier to manage and optimize at each stage. This approach also provides flexibility to adjust plans based on performance and market conditions.

Testing and optimization should be integral parts of each expansion phase. After setting up new rigs, thoroughly test their performance and stability. Make necessary adjustments to ensure optimal operation before proceeding with further expansion. This iterative process helps in identifying and resolving issues early, preventing them from escalating as the operation grows.

Infrastructure Enhancement:

Expanding your mining operation requires significant upgrades to your infrastructure. Start by ensuring adequate physical space for additional mining rigs, considering future expansions as well. The layout should facilitate easy access for maintenance and provide efficient airflow to manage heat dissipation.

Upgrading the electrical infrastructure is crucial to handle increased power demands. This may involve installing additional circuits, transformers, and power distribution units (PDUs). Ensuring a stable and reliable power supply is essential to prevent downtime and protect your hardware from damage due to power fluctuations.

Cooling systems are another critical component of the infrastructure. As you add more rigs, the heat generated increases, necessitating enhanced cooling solutions. Upgrading existing cooling systems or implementing advanced technologies like liquid cooling or immersion cooling can effectively manage the increased heat output. Efficient cooling not only protects your hardware but also enhances its performance and longevity.

Financial Management:

Scaling up involves significant financial investment. Therefore, effective financial management is essential to ensure that the expansion is economically viable. Start by allocating a budget that covers all aspects of scaling, including hardware procurement, infrastructure upgrades, and ongoing operational expenses. Having a well-defined budget helps in managing costs and avoiding overspending.

Cost management strategies are crucial to control expenses and maximize profitability. This includes negotiating better rates for bulk hardware purchases, optimizing energy consumption, and seeking out cost-effective solutions for infrastructure enhancements. Regular financial reviews and adjustments based on performance data help in maintaining financial health and achieving a good return on investment (ROI).

Managing Multiple Rigs

Centralized Management Platforms:

As the number of mining rigs increases, managing them individually becomes impractical. Centralized management platforms, such as Hive OS, Awesome Miner, or Minerstat, provide a solution by allowing you to monitor and control multiple rigs from a single interface. These platforms offer comprehensive tools for tracking performance metrics, managing settings, and automating routine tasks.

Automation tools are particularly valuable in large-scale operations. They can streamline tasks like firmware updates, rebooting rigs, and adjusting settings, reducing the need for manual intervention. This not only saves time but also minimizes the risk of human error, enhancing the overall efficiency of the operation.

Network Configuration:

Ensuring stable and high-speed connectivity is essential for efficient mining operations. Wired connections (Ethernet) are preferable to wireless connections, as they provide more reliable network performance and reduce the risk of interruptions.

IP address management is another important aspect of network configuration. Assign static IP addresses to each rig for easier management and troubleshooting. This helps in quickly identifying and resolving network issues, ensuring minimal downtime and optimal performance.

Load Balancing:

Distributing the workload evenly across all mining rigs is crucial to prevent overloading and ensure consistent performance. Implementing load balancing strategies helps in optimizing the use of available resources and maintaining the stability of the operation.

Redundancy measures are also important to maintain continuity in case of hardware failures or network issues. This includes having backup systems and failover mechanisms in place to quickly switch to alternative resources when needed, minimizing downtime and maintaining productivity.

Performance Monitoring:

Continuous monitoring of performance metrics is essential to ensure the optimal operation of mining rigs. Use monitoring tools to track key metrics such as hash rate, temperature, power consumption, and uptime. Real-time monitoring helps in identifying and addressing issues promptly, preventing performance degradation and hardware damage.

Setting up alerts and notifications for critical events, such as hardware failures, temperature spikes, and network issues, ensures that you can respond quickly to problems. This proactive approach to monitoring helps in maintaining high efficiency and preventing costly downtime.

Optimizing Power and Cooling

Power Optimization:

Energy efficiency is a key factor in the profitability of mining operations. Investing in energy-efficient hardware can significantly reduce power consumption and operational costs. High-quality power distribution units (PDUs) ensure stable power delivery to all rigs, preventing overloads and protecting hardware from power fluctuations.

Implementing voltage regulation solutions can further enhance power optimization. Voltage regulators protect hardware from power surges and fluctuations, ensuring consistent power delivery and reducing the risk of damage. This not only extends the lifespan of the hardware but also improves its performance and reliability.

Cooling Optimization:

Efficient cooling is essential to manage the substantial heat generated by mining rigs. Enhanced air-cooling systems, such as high-performance fans and optimized airflow designs, can effectively dissipate heat and maintain safe operating temperatures.

Liquid cooling and immersion cooling solutions offer even greater efficiency in managing heat. Liquid cooling uses specialized coolant fluids to absorb and transport heat away from the hardware, while immersion cooling involves submerging the hardware in a thermally conductive, but electrically insulating liquid. These advanced cooling solutions can significantly reduce energy costs associated with traditional air cooling and enhance the performance and lifespan of the hardware.

Environmental controls are also important for maintaining optimal operating conditions. Monitoring and controlling temperature, humidity, and airflow in the mining facility helps in preventing overheating and ensuring consistent performance.

Energy Efficiency:

Conducting regular energy audits helps in identifying areas for improvement in energy efficiency. This involves analyzing energy consumption patterns and implementing measures to reduce waste and optimize usage. Energy-efficient hardware, optimized cooling solutions, and renewable energy sources can all contribute to reducing operational costs and environmental impact.

Exploring renewable energy sources, such as solar, wind, or hydroelectric power, can provide cost-effective and sustainable energy solutions for mining operations. Integrating renewable energy not only reduces reliance on fossil fuels but also enhances the sustainability and public perception of the mining operation.

Handling Increased Maintenance and Monitoring

Maintenance Protocols:

Regular maintenance is crucial to ensure the optimal performance and longevity of mining hardware. Implement a scheduled maintenance plan that includes routine checks, cleaning, firmware updates, and thermal paste reapplication. Preventive maintenance helps in identifying and addressing potential issues before they lead to hardware failures or performance degradation.

Monitoring Systems:

Comprehensive monitoring systems provide real-time data on the performance and health of mining rigs. Key metrics to track include hash rate, temperature, power consumption, and error rates. Analyzing this data helps in identifying trends, optimizing performance, and predicting potential issues.

Data analytics tools can provide valuable insights into the performance of the mining operation. By analyzing performance data, miners can identify patterns, optimize settings, and make informed decisions to enhance efficiency and profitability.

Support and Troubleshooting:

Access to technical support is essential for resolving hardware and software issues. This may involve having in-house expertise or partnering with external service providers. Developing troubleshooting protocols helps in quickly diagnosing and resolving issues, minimizing downtime and maintaining productivity.

Documentation and Training:

Standard Operating Procedures (SOPs) are essential for ensuring consistency and efficiency in maintenance and monitoring tasks. Develop and document SOPs for all tasks, and ensure that all team members are familiar with these procedures. Providing training programs for staff enhances their skills and knowledge, enabling them to effectively manage and maintain the mining operation.

Scaling Bitcoin mining operations requires careful planning, strategic investments, and ongoing optimization to achieve success. By implementing effective strategies for scaling up, managing multiple rigs, optimizing power and cooling, and handling increased maintenance and monitoring, miners can expand their operations efficiently and sustainably.

Adopting centralized management platforms, enhancing infrastructure, and focusing on energy efficiency are key components of successful scaling. Continuous monitoring, regular maintenance, and comprehensive training programs ensure that the expanded mining operation runs smoothly and remains profitable.

As the cryptocurrency mining landscape evolves, staying informed about the latest technologies and best practices will be crucial for maintaining a competitive edge and achieving long-term success in large-scale mining operations. By embracing innovation and sustainability, miners can position themselves for future growth and profitability in the dynamic world of cryptocurrency mining.

Chapter 17:
Mining Future Forecast

The future of cryptocurrency mining is influenced by various trends, technological advancements, regulatory impacts, and emerging innovations. This chapter delves into these factors to provide a comprehensive forecast of the mining industry's trajectory, offering insights into what miners can expect and how they can prepare for the changes ahead.

Trends in Cryptocurrency Mining

Increasing Difficulty and Hash Rate:

As the popularity and value of cryptocurrencies like Bitcoin continue to rise, more miners are joining the network, leading to an increase in the overall hash rate. This metric measures the total computational power being used to mine and process transactions on the blockchain. The higher the hash rate, the more secure the network becomes, but it also increases the difficulty of mining. The network automatically adjusts the difficulty level to ensure that blocks are mined at a consistent rate, roughly every ten minutes for Bitcoin.

For individual miners, this increasing difficulty means that more advanced and powerful hardware is required to stay competitive. Older models of ASICs (Application-Specific Integrated Circuits) become obsolete faster, pushing miners to constantly upgrade to newer, more efficient models. This trend emphasizes the need for significant capital investment and continual technological advancement to maintain profitability.

Geographic Shifts:

The geographical landscape of mining is shifting due to various factors such as regulatory changes, energy costs, and the availability of infrastructure. For instance, China's crackdown on cryptocurrency mining led to a massive migration of mining operations to other countries with more favorable regulations and lower energy costs. The United States, Canada, Kazakhstan, and Russia have become prominent destinations for these migrating miners.

Regions with abundant renewable energy sources are particularly attractive. Iceland and Canada, for example, offer geothermal and hydroelectric power, respectively, providing environmentally friendly and cost-effective energy solutions for mining operations. This shift not only influences the global distribution of mining power but also impacts local economies and energy grids in the host countries.

Environmental Concerns:

The environmental impact of Bitcoin mining has become a major concern. The process is energy-intensive, and much of this energy has historically come from fossil fuels, leading to significant carbon emissions. Public awareness and criticism of the carbon footprint of mining have grown, prompting miners to seek more sustainable practices.

To address these concerns, many mining operations are now investing in renewable energy sources. Solar, wind, and hydroelectric power are being increasingly integrated into mining setups to reduce reliance on fossil fuels. Additionally, some companies are participating in carbon offset programs or investing in carbon capture technologies to mitigate their environmental impact. These efforts are not only beneficial for the environment but also improve the public perception of cryptocurrency mining.

Decentralization and Pool Dominance:

Mining pools have become a dominant force in the cryptocurrency mining landscape. By combining the hash power of multiple miners, pools increase the chances of solving a block and earning rewards, which are then distributed among the participants. However, this trend towards pool dominance has raised concerns about centralization. When a few large pools control a significant portion of the network's hash rate, the risk of a 51% attack—where a single entity could potentially manipulate the blockchain—increases.

To counter this, there are ongoing efforts to promote decentralization. Some initiatives focus on encouraging solo mining by developing more efficient algorithms and hardware that make it feasible for individual miners to participate without relying on pools. Other strategies involve the creation of decentralized mining pools that distribute control and rewards more evenly among participants.

Technological Advancements in Mining

Hardware Evolution:

The development of mining hardware has been a continuous race for efficiency and power. ASICs, which are specialized chips designed specifically for mining, have revolutionized the industry by providing significantly higher hash rates compared to traditional GPUs (Graphics Processing Units). The evolution of ASICs continues, with each new generation offering better performance and energy efficiency.

One of the emerging areas of interest is quantum computing. Although still in its infancy, quantum computing has the potential to disrupt current cryptographic algorithms used in mining. Quantum computers could theoretically solve cryptographic puzzles much faster than classical computers, posing both a challenge and an opportunity for the mining industry. While the widespread adoption of quantum computing is likely years away, it remains a critical area of research and development.

Energy Efficiency:

Energy efficiency is a key factor in the profitability and sustainability of mining operations. Advanced cooling solutions, such as immersion cooling and liquid cooling, are becoming more common. These technologies allow miners to maintain optimal operating temperatures, reduce energy consumption, and extend the lifespan of their hardware.

Immersion cooling involves submerging mining hardware in a thermally conductive, but electrically insulating liquid. This method effectively dissipates heat and can significantly reduce energy costs associated with traditional air cooling. Liquid cooling, on the other hand, uses specialized coolant fluids to absorb and transport heat away from the hardware. Both methods represent significant advancements in maintaining high performance while managing the substantial heat generated by mining rigs.

Software and Algorithms:

Mining software continues to evolve, with ongoing improvements aimed at enhancing efficiency and performance. Optimized mining algorithms reduce power consumption and increase hash rates, allowing miners to maximize their hardware's potential. The integration of AI and machine learning is also becoming more prevalent, enabling smarter and more adaptive mining operations.

AI can predict hardware failures, optimize settings dynamically based on real-time data, and even automate maintenance tasks. Machine learning algorithms analyze vast amounts of performance data to identify patterns and optimize mining strategies, further improving efficiency and profitability.

Blockchain Innovations:

Blockchain technology itself is continuously evolving. Layer 2 solutions, such as the Lightning Network for Bitcoin, aim to improve transaction speed and reduce fees, potentially altering the mining incentive structure. These solutions enable off-chain transactions, which are faster and cheaper, while still maintaining the security of the main blockchain.

Additionally, the shift from Proof of Work (PoW) to Proof of Stake (PoS) in some cryptocurrencies, like Ethereum, represents a significant change. PoS is less energy-intensive, as it does not require the same level of computational power as PoW. This transition impacts hardware manufacturers and mining operations, as the demand for traditional mining hardware decreases.

Impact of Regulations and Market Conditions

Regulatory Environment:

Government regulations are a major factor influencing the future of cryptocurrency mining. Regulatory actions can range from supportive measures, such as tax incentives for using renewable energy, to restrictive policies, like outright bans on mining activities. For instance, China's ban on cryptocurrency mining led to a significant reshuffling of the global mining landscape, forcing miners to relocate to more favorable jurisdictions.

In the United States, regulatory clarity is gradually improving, with some states actively encouraging mining through favorable policies and access to renewable energy. However, increased scrutiny is also expected, with potential requirements for environmental impact assessments and financial reporting. Navigating this regulatory landscape requires miners to stay informed and adaptable.

Market Conditions:

The volatility of cryptocurrency prices directly impacts mining profitability. During periods of high prices, mining activity intensifies as miners seek to maximize returns. Conversely, prolonged downturns can force miners to shut down operations or scale back to reduce losses. Energy prices also play a crucial role, as electricity is a major operational cost. Regions with stable and low-cost energy sources are more attractive for mining operations.

Market conditions are influenced by various factors, including technological advancements, macroeconomic trends, and geopolitical events. Miners must continuously adapt to these changing conditions to maintain profitability and competitiveness.

Geopolitical Factors:

Geopolitical stability and international trade policies can significantly affect the mining industry. Trade policies, such as tariffs on hardware components, can impact the cost and availability of mining equipment. Political stability in regions hosting mining operations influences the long-term feasibility and security of these investments.

For example, the political and economic stability in North America has made it an attractive destination for miners relocating from regions with less favorable conditions. Conversely, political instability or unfavorable trade policies in other regions can disrupt mining activities and supply chains.

Predictions for the Future of Mining

Consolidation and Professionalization:

The cryptocurrency mining industry is likely to see further consolidation, with larger, more professional operations dominating the landscape. The economies of scale enjoyed by large mining farms allow them to operate more efficiently and profitably than smaller, individual miners. This trend may lead to fewer, but larger and more sophisticated mining entities.

Professionalization of mining operations includes adopting advanced technologies, implementing best practices, and maintaining rigorous operational standards. This shift towards professionalization will enhance the overall efficiency and reliability of mining operations, making them more resilient to market fluctuations and regulatory changes.

Sustainable Mining Practices:

Sustainability will be a key focus for the future of mining. The push towards green mining practices will drive the adoption of renewable energy sources and energy-efficient technologies. Operations powered by solar, wind, and hydroelectric energy will become more common, reducing the environmental impact of mining.

Miners will aim for carbon neutrality through the use of renewable energy, carbon offsets, and improved energy management practices. These efforts will not only reduce the carbon footprint of mining operations but also enhance their public image and regulatory compliance.

Decentralization Efforts:

Efforts to decentralize mining and reduce the dominance of large pools will continue. Decentralization is crucial for maintaining the security and integrity of blockchain networks. Initiatives to promote solo mining and the development of decentralized mining pools will gain traction, fostering a more resilient and distributed network.

Community-driven initiatives will support decentralized mining, encouraging participation from smaller miners and hobbyists. This movement towards decentralization will help ensure that the control of mining power remains distributed, preventing centralization risks.

Technological Integration:

The integration of AI and IoT technologies will revolutionize mining operations. AI-driven analytics and machine learning will optimize performance, predict maintenance needs, and adjust settings dynamically. IoT devices equipped with smart sensors will monitor environmental conditions, hardware performance, and energy consumption, providing valuable data for optimizing operations.

Blockchain innovations will continue to shape the mining landscape. Improved consensus mechanisms, such as hybrid models combining PoW and PoS, will enhance network security and efficiency. Advancements in blockchain interoperability will enable more efficient cross-chain transactions and collaborations, potentially altering mining incentives and rewards.

Emerging Technologies: AI, IoT, and Blockchain Innovations

Artificial Intelligence (AI):

AI has the potential to transform mining operations by providing predictive maintenance capabilities and optimization algorithms. Predictive maintenance uses AI to analyze historical data and predict hardware failures, allowing miners to perform maintenance before issues arise. This reduces downtime and extends the lifespan of mining equipment.

Optimization algorithms driven by AI can dynamically adjust mining parameters based on real-time data, maximizing efficiency and profitability. These algorithms can analyze factors such as energy prices, network difficulty, and hardware performance to make informed decisions about mining operations.

Internet of Things (IoT):

The integration of IoT devices into mining operations enhances monitoring and automation capabilities. Smart sensors can continuously monitor environmental conditions, hardware performance, and energy consumption, providing real-time data for optimizing operations.

IoT-enabled automation can streamline various aspects of mining operations, from adjusting cooling systems to managing power distribution. This level of automation improves efficiency, reduces operational costs, and minimizes the risk of human error.

Blockchain Innovations:

Blockchain technology itself is evolving, with innovations that significantly impact mining operations. Improved consensus mechanisms, such as Proof of Stake (PoS) and hybrid models, offer energy-efficient alternatives to traditional Proof of Work (PoW) mining. These mechanisms reduce the energy consumption and environmental impact of mining while maintaining network security.

A notable example of this evolution is Ethereum, which successfully migrated from PoW to PoS in 2022, showcasing the potential for major blockchain networks to adopt more sustainable and efficient consensus mechanisms.

Advancements in blockchain interoperability enable more efficient cross-chain transactions and collaborations. This can alter mining incentives and rewards, encouraging miners to participate in multiple blockchain networks and diversify their operations.

The future of cryptocurrency mining is poised for significant changes driven by technological advancements, regulatory developments, market conditions, and emerging innovations. As the industry evolves, miners must adapt to these changes by embracing new technologies, optimizing operations for efficiency and sustainability, and navigating the complex regulatory landscape.

The trends and predictions outlined in this chapter highlight the importance of staying informed and agile in the face of a rapidly changing mining environment. By leveraging emerging technologies like AI and IoT, adopting sustainable practices, and staying ahead of regulatory changes, miners can position themselves for long-term success in the competitive and dynamic world of cryptocurrency mining.

Chapter 18:
Economic Aspects of Mining

Understanding the economic aspects of cryptocurrency mining is essential for achieving profitability and long-term success. This chapter delves into the fundamental economic principles behind cryptocurrency, the dynamics of supply and demand, the impact of mining on coin value, market cycles, profitability analysis, operational costs, return on investment (ROI) calculations, and economic analysis including 'hashprice' management.

Economic Principles Behind Cryptocurrency

Cryptocurrencies operate on economic principles similar to traditional assets, but with unique characteristics driven by blockchain technology and decentralization.

Scarcity and Value: Cryptocurrencies like Bitcoin have a finite supply, with Bitcoin capped at 21 million coins. This scarcity principle is akin to precious metals like gold, where limited supply supports value. As the supply diminishes, the value tends to increase, assuming constant or rising demand.

Decentralization and Trust: Unlike fiat currencies, cryptocurrencies are decentralized and operate without central banks or governments. Trust in the network is established through cryptographic protocols and consensus mechanisms, making the economic model heavily reliant on technological integrity and community consensus.

Network Effects: The value of a cryptocurrency often benefits from network effects. As more users, developers, and businesses adopt a particular cryptocurrency, its utility and value can increase. This self-reinforcing mechanism is crucial for the growth and stability of the cryptocurrency ecosystem.

Supply and Demand Dynamics

The supply and demand dynamics of cryptocurrencies are influenced by several factors:

Fixed Supply: For cryptocurrencies like Bitcoin, the fixed supply cap ensures that the total number of coins will never exceed a predetermined limit. This creates a predictable supply curve, contrasting with fiat currencies where supply can be adjusted by central banks.

Demand Drivers: Demand for cryptocurrencies can be driven by various factors, including:

- **Speculative Investment:** Investors seeking high returns drive demand, often influenced by market sentiment and media coverage.
- **Utility and Adoption:** The real-world utility of a cryptocurrency, such as its use in transactions, smart contracts, or decentralized applications (dApps), can drive demand.
- **Regulatory Environment:** Favorable regulations can boost demand by providing legal clarity and security, while unfavorable regulations can suppress demand.

Market Liquidity: Liquidity in the cryptocurrency market affects the ease with which assets can be bought or sold without significantly impacting their price. Higher liquidity generally leads to more stable prices, while low liquidity can result in high volatility.

Impact of Mining on Coin Value

Mining plays a crucial role in the cryptocurrency ecosystem, influencing the value of coins through several mechanisms:

Coin Supply: Mining introduces new coins into circulation. The rate at which new coins are mined affects the overall supply and can influence price dynamics. For Bitcoin, the block reward halving event every four years reduces the rate of new coin creation, often leading to price increases as supply growth slows.

Network Security: The security of a cryptocurrency network is directly tied to its hash rate. A higher hash rate makes the network more secure against attacks, enhancing trust and stability. This security can positively impact the coin's value as investors and users have greater confidence in the network's integrity.

Transaction Processing: Miners validate and process transactions, ensuring the smooth operation of the network. Efficient and reliable transaction processing can enhance the user experience and utility of cryptocurrency, potentially increasing its value.

Market Cycles and Their Effects on Mining

Cryptocurrency markets are known for their volatility and cyclical nature. Understanding these market cycles is crucial for miners to navigate profitability and operational strategies.

Bull Markets: During bull markets, cryptocurrency prices rise significantly, driven by increased demand and positive market sentiment. High prices lead to increased mining profitability as the value of mined coins rises. This often results in a surge of new miners entering the market, increasing competition and hash rate.

Bear Markets: Bear markets are characterized by declining prices and reduced market activity. Mining profitability can decrease as the value of mined coins falls. During these periods, less efficient miners may be forced to shut down, leading to a decline in hash rate. However, bear markets also offer opportunities for miners to acquire hardware at lower costs and prepare for the next bull cycle.

Market Corrections: Market corrections are short-term price declines within a longer-term trend. These corrections can impact mining profitability and operational decisions. Miners must remain agile, adjusting strategies to maintain profitability during these volatile periods.

Analyzing Mining Profitability

Analyzing the profitability of mining operations involves evaluating various factors and metrics:

Hash Rate: The hash rate of mining hardware directly impacts the number of coins mined. Higher hash rates increase the probability of solving blocks and earning rewards, enhancing profitability.

Energy Costs: Energy consumption is a significant operational cost for miners. The efficiency of mining hardware, measured in watts per hash (W/TH), directly affects energy costs. Lower energy costs improve profitability.

Hardware Costs: The initial investment in mining hardware and ongoing maintenance costs are critical factors in profitability analysis. Miners must balance the cost of acquiring and maintaining hardware with the expected returns from mining operations.

Difficulty Adjustment: The mining difficulty adjusts periodically based on the network's total hash rate. Higher difficulty levels require more computational power to solve blocks, impacting the profitability of mining operations.

Cryptocurrency Prices: The market price of the mined cryptocurrency is a major determinant of profitability. Fluctuations in price can significantly impact the value of mining rewards and overall profitability.

Pool Fees: For miners participating in mining pools, fees charged by the pool operators can affect profitability. Lower fees can improve net returns, while higher fees reduce the effective mining rewards.

Operational Costs in Mining

Operational costs in mining encompass several key areas:

Electricity: Electricity is the most significant ongoing cost for mining operations. The price per kilowatt-hour (kWh) varies by region and can significantly impact overall profitability. Miners often seek locations with lower electricity rates to reduce costs.

Cooling: Cooling systems are essential to manage the heat generated by mining hardware. The cost of cooling solutions, including air conditioning, liquid cooling, and immersion cooling, adds to operational expenses.

Maintenance: Regular maintenance is necessary to ensure optimal performance and longevity of mining hardware. This includes cleaning, thermal paste reapplication, and hardware repairs or replacements.

Facility Costs: The costs associated with maintaining the physical infrastructure, such as rent, security, and insurance, contribute to operational expenses. Optimizing facility management can help reduce these costs.

Personnel: Labor costs for managing and maintaining mining operations are another consideration. Skilled technicians are required to handle hardware setup, troubleshooting, and maintenance tasks.

ROI Calculation for Mining Investments

Calculating the return on investment (ROI) for mining operations involves evaluating the initial capital expenditure and ongoing operational costs against the expected returns.

Initial Investment: The initial investment includes the cost of mining hardware, infrastructure setup, and any associated expenses. This capital outlay forms the basis for ROI calculations.

Revenue Projections: Revenue projections are based on the expected mining rewards over a given period. This includes the number of coins mined and their market value. Projections should account for factors such as mining difficulty adjustments and price fluctuations.

Payback Period: The payback period is the time required to recoup the initial investment from mining revenues. A shorter payback period indicates a more favorable investment.

Net Profit: Net profit is calculated by subtracting total operational costs from total mining revenue. This metric provides a clear picture of the profitability of mining operations.

ROI Formula:

$$ROI = \left(\frac{\text{Net Profit}}{\text{Initial Investment}}\right) \times 100$$

The ROI formula helps in determining the percentage return on the initial investment, allowing miners to compare the profitability of different mining ventures.

Economic Analysis and 'Hashprice' Management

'Hashprice' is a critical metric in mining economics, representing the revenue generated per terahash per second (TH/s) of computational power.

Calculating Hashprice: Hashprice is calculated by dividing the total revenue earned from mining by the total hash rate of the operation. This metric helps in evaluating the efficiency and profitability of mining hardware.

Monitoring Hashprice: Regular monitoring of hashprice is essential for making informed operational decisions. Fluctuations in hashprice can indicate changes in market conditions, mining difficulty, and network hash rate.

Adjusting Operations: Based on hashprice analysis, miners can adjust their operations to optimize profitability. This may involve reconfiguring hardware, adjusting energy consumption, or switching mining pools.

Strategic Decision-Making: Hashprice management is a strategic tool for decision-making. By analyzing hashprice trends, miners can make informed choices about hardware upgrades, energy contracts, and scaling operations.

Understanding the economic aspects of mining is crucial for navigating the complexities of cryptocurrency mining operations. By grasping the fundamental economic principles, supply and demand dynamics, and the impact of mining on coin value, miners can make informed decisions to enhance profitability.

Market cycles, operational costs, and ROI calculations are essential components of a comprehensive economic analysis. Effective hashprice management and continuous monitoring of key metrics help miners adapt to changing market conditions and maintain a competitive edge.

By integrating these economic principles and strategies, miners can optimize their operations, achieve sustainable profitability, and position themselves for long-term success in the dynamic and evolving world of cryptocurrency mining.

Chapter 19:
Legal Considerations

Cryptocurrency mining, while lucrative, operates within a complex and evolving legal landscape. Navigating the legal aspects of mining involves understanding regulatory environments, compliance requirements, and the potential risks associated with changing laws. This chapter delves into these areas, providing insights into the regulatory risks to Bitcoin mining, compliance strategies, tax implications, and case studies illustrating the impact of regulations on mining operations.

Legal Aspects of Cryptocurrency Mining

Understanding the Legal Landscape:

Cryptocurrency mining exists in a legal gray area in many jurisdictions, primarily because the regulatory frameworks have not yet caught up with the rapid development of the technology. The legal status of mining can vary widely from one country to another, and even within different regions of the same country.

Key Legal Issues:

- **Property Rights:** Miners must understand the property rights associated with the mined cryptocurrencies. In some jurisdictions, cryptocurrencies are considered legal property, while in others, they are not.
- **Contract Law:** Mining operations often involve complex contracts, especially when engaging with third-party hosting services or mining pools. Ensuring these contracts are legally sound is crucial.

- **Environmental Regulations:** Mining operations must comply with local environmental laws, particularly those related to energy consumption and carbon emissions.

Regulatory Environment and Compliance

The regulatory environment for cryptocurrency mining is dynamic and varies significantly across different regions. Understanding and complying with these regulations is critical to avoid legal issues and ensure the longevity of mining operations.

Global Regulatory Overview:

- **United States:** The U.S. regulatory landscape is fragmented, with different states adopting varying approaches to cryptocurrency mining. States like Texas and Wyoming are known for their favorable regulations, while others have imposed restrictions.
- **China:** Historically a major hub for Bitcoin mining, China has implemented stringent regulations and outright bans on mining activities, causing a significant shift in the global mining landscape.
- **European Union:** The EU has been proactive in regulating cryptocurrencies, focusing on anti-money laundering (AML) and environmental impacts. Countries like Iceland and Norway, with abundant renewable energy, remain attractive for miners.
- **Other Regions:** Countries like Canada and Kazakhstan have emerged as key players in the mining industry, offering regulatory clarity and favorable conditions for miners.

Key Regulatory Compliance Areas:

- **Licensing and Permits:** Depending on the jurisdiction, mining operations may require specific licenses or permits to operate legally. This includes business licenses, environmental permits, and energy usage approvals.

- **AML and KYC Requirements:** Compliance with AML and Know Your Customer (KYC) regulations is crucial to prevent illegal activities such as money laundering and terrorism financing. Mining pools and service providers must implement robust AML/KYC procedures.
- **Energy Regulations:** Mining operations must adhere to local energy regulations, including those related to energy consumption, efficiency, and carbon emissions. This is especially important in regions with strict environmental laws.

Regulatory Risks to Bitcoin Mining

Overview of Regulatory Risks:

Regulatory risks refer to the uncertainties and potential adverse effects that changing laws and regulations can have on mining operations. These risks can impact profitability, operational stability, and even the legal viability of mining activities.

- **Legal Uncertainty:** The lack of clear and consistent regulations creates uncertainty for miners, making it difficult to plan and operate long-term.
- **Sudden Regulatory Changes:** Governments can introduce new regulations or bans with little notice, significantly impacting mining operations.
- **Environmental Regulations:** Increasing scrutiny on the environmental impact of mining can lead to stricter regulations on energy usage and carbon emissions.

Impact of Local and International Regulations:

- **Local Regulations:** Local laws can directly affect mining operations, including zoning laws, noise regulations, and energy usage limits. Compliance with these regulations is essential to avoid fines and operational disruptions.

- **International Regulations:** International regulations, such as sanctions and cross-border data flow restrictions, can impact the global operations of mining companies. Understanding these regulations is crucial for miners operating in multiple countries.

Compliance Strategies:

- **Proactive Compliance:** Staying ahead of regulatory changes by proactively engaging with regulators and participating in industry associations can help miners shape favorable regulations and ensure compliance.
- **Legal Counsel:** Engaging with legal experts who specialize in cryptocurrency and mining law can provide valuable guidance and help navigate complex legal landscapes.
- **Environmental Sustainability:** Adopting environmentally sustainable practices and technologies can help mitigate the risks associated with environmental regulations and improve public perception.

Case Studies of Regulatory Impacts on Mining Operations:

- **China's Ban on Mining:** China's ban on cryptocurrency mining in 2021 had a profound impact on the global mining industry. Many miners relocated to other countries, significantly shifting the geographic distribution of mining power.
- **Kazakhstan's Regulatory Environment:** Kazakhstan emerged as a key destination for displaced miners from China due to its favorable regulations and abundant energy resources. However, political instability and subsequent regulatory changes posed new challenges for miners.
- **United States State Regulations:** Different states in the U.S. have adopted varied approaches to mining regulation.

For example, New York imposed a moratorium on new mining operations pending an environmental review, while Texas has embraced mining with open arms, offering incentives and access to renewable energy.

Tax Implications of Mining

Mining activities have significant tax implications that miners must understand to ensure compliance and optimize their tax liabilities.

Income Tax:

Mining rewards are typically considered taxable income. Miners must report the fair market value of the mined coins at the time they are received as income. This applies to both individual and corporate miners.

Capital Gains Tax:

When mined coins are sold or exchanged, they may be subject to capital gains tax. The taxable gain is calculated based on the difference between the sale price and the fair market value of the coins at the time of mining. Understanding the tax treatment of these transactions is crucial for accurate reporting.

Deductible Expenses:

Mining operations incur various expenses that may be deductible for tax purposes. These include the cost of mining hardware, electricity, cooling, rent, and maintenance. Keeping detailed records of these expenses is essential for maximizing deductions and minimizing tax liabilities.

Tax Compliance:

- **Record Keeping:** Maintaining detailed records of all mining activities, income, and expenses is essential for accurate tax reporting and compliance.

- **Tax Planning:** Engaging with tax professionals who understand cryptocurrency taxation can help miners optimize their tax strategies and ensure compliance with local tax laws.

Staying Informed and Compliant with Laws

The legal and regulatory landscape for cryptocurrency mining is constantly evolving. Staying informed and compliant with these laws is crucial for the sustainability and legality of mining operations.

Continuous Monitoring:

Regularly monitor regulatory developments at local, national, and international levels. Subscribing to industry newsletters, participating in forums, and engaging with legal experts can provide valuable insights into regulatory changes.

Industry Associations:

Joining industry associations and advocacy groups can help miners stay informed about regulatory trends and participate in discussions that shape future regulations. These organizations often provide resources, updates, and networking opportunities.

Legal Audits:

Conducting regular legal audits of mining operations can help identify potential compliance issues and ensure adherence to current laws. Engaging with legal counsel for periodic reviews can mitigate risks and ensure ongoing compliance.

Public Engagement:

Engaging with the public and policymakers to educate them about the benefits and challenges of cryptocurrency mining can help build positive relationships and influence regulatory outcomes.

Transparency and proactive communication can improve the public perception of mining operations.

Navigating the legal considerations of cryptocurrency mining is essential for maintaining compliance, managing risks, and ensuring the sustainability of operations. Understanding the regulatory environment, addressing regulatory risks, and implementing effective compliance strategies are crucial for success in the dynamic world of cryptocurrency mining.

By staying informed about regulatory developments, engaging with industry associations, and adopting proactive compliance measures, miners can navigate the complex legal landscape and position themselves for long-term success. As the industry continues to evolve, a thorough understanding of legal considerations will remain a cornerstone of successful and sustainable mining operations.

The Blockchain Academy LLC

Chapter 20:
Environmental Sustainability and Compliance

As the cryptocurrency mining industry continues to grow, so does the scrutiny over its environmental impact. Bitcoin mining, in particular, has garnered significant attention for its high energy consumption and associated carbon emissions. This chapter delves into the various aspects of environmental sustainability and compliance, offering a comprehensive guide on assessing environmental impacts, reducing energy consumption, adhering to legal standards, engaging in sustainability certifications, and implementing renewable energy solutions.

Assessing the Environmental Impact of Mining

Before addressing the sustainability of a mining operation, it's crucial to understand its environmental footprint. This involves a thorough assessment of energy consumption, carbon emissions, and other ecological impacts.

Energy Consumption: Cryptocurrency mining, especially Bitcoin, requires substantial computational power, which translates into high energy consumption. Mining rigs, particularly ASICs, are notorious for their energy demands. To quantify this, miners should monitor the kilowatt-hours (kWh) consumed by their equipment, including both the mining hardware and the cooling systems required to maintain optimal temperatures.

Carbon Footprint: The carbon footprint of a mining operation is largely determined by the source of its electricity. Energy derived from fossil fuels, such as coal and natural gas, results in higher carbon emissions compared to renewable sources like solar, wind, or hydroelectric power. Calculating the total carbon emissions involves analyzing the energy mix of the power grid and applying emissions factors for different energy sources. Understanding this footprint is essential for developing strategies to mitigate environmental impact.

E-Waste Generation: Mining hardware has a limited lifespan and contributes to electronic waste (e-waste) when it becomes obsolete. The rapid pace of technological advancement in mining equipment exacerbates this issue, leading to significant volumes of e-waste. Assessing the volume of e-waste generated and implementing proper disposal and recycling practices are critical steps in minimizing the environmental impact of outdated equipment.

Strategies for Reducing Energy Consumption

Reducing energy consumption not only lowers operational costs but also mitigates the environmental impact of mining operations. Here are several strategies to achieve this:

Energy-Efficient Hardware: Investing in the latest, most energy-efficient mining hardware can drastically reduce electricity consumption. Modern ASICs are designed to deliver higher hash rates while consuming less power. Regularly upgrading to newer models ensures that mining operations remain competitive and environmentally sustainable.

Optimized Cooling Solutions: Cooling is a significant part of a mining operation's energy consumption. Implementing advanced cooling solutions can improve energy efficiency:

- **Air Cooling:** Enhance airflow within the mining facility by using high-efficiency fans and optimizing the layout to facilitate natural ventilation.
- **Liquid Cooling:** Liquid cooling systems, which circulate a coolant around the hardware, can remove heat more effectively than air cooling.

- **Immersion Cooling:** This innovative technique involves submerging mining hardware in a thermally conductive but electrically insulating liquid, providing superior cooling efficiency and reducing energy usage.

Energy Management Systems: Implementing energy management systems allows for real-time monitoring and optimization of energy consumption. These systems can identify inefficiencies and suggest adjustments to improve energy use. Automated controls can dynamically adjust power settings based on operational needs, further enhancing efficiency.

Demand Response Programs: Participating in demand response programs helps balance the load on the electricity grid. During peak demand periods, mining operations can reduce or shift their energy usage, contributing to grid stability and potentially earning financial incentives. This not only supports the local power infrastructure but also promotes a positive relationship with energy providers.

Legal Frameworks and Compliance with Environmental Standards

Navigating the legal landscape of environmental regulations is crucial for the sustainability and legality of mining operations. Different regions have varying laws and standards that miners must comply with.

International Regulations: Global agreements such as the Paris Agreement aim to combat climate change by reducing carbon emissions. Countries that are signatories to such agreements may implement national policies to meet their commitments, impacting mining operations within their borders.

National and Local Regulations: Environmental regulations can vary significantly between countries and even within regions of the same country. Key areas of regulation include:

- **Emission Limits:** Laws may set limits on greenhouse gas emissions, requiring miners to reduce their carbon output.
- **Energy Efficiency Standards:** Some regions mandate minimum energy efficiency standards for industrial operations, including mining.
- **Waste Management Laws:** Proper disposal and recycling of e-waste are often regulated to minimize environmental impact.

Compliance Strategies:

- **Regular Audits:** Conducting regular environmental audits helps ensure compliance with applicable laws and standards. These audits can identify areas for improvement and track progress towards sustainability goals.
- **Legal Counsel:** Engaging with legal experts who specialize in environmental law provides guidance on compliance and helps navigate complex regulations.
- **Environmental Impact Assessments (EIA):** Performing EIAs before setting up or expanding mining operations can identify potential environmental risks and mitigation measures.

Engaging with Sustainability Certifications and Audits

Achieving sustainability certifications and undergoing regular audits demonstrate a mining operation's commitment to environmental responsibility.

Certifications:

- **ISO 14001:** This international standard sets out criteria for an effective environmental management system, helping organizations improve their environmental performance through more efficient use of resources and reduction of waste.

- **LEED Certification:** The Leadership in Energy and Environmental Design (LEED) certification is a widely used green building rating system. Mining facilities can pursue LEED certification to showcase their commitment to sustainability and energy efficiency.

Sustainability Audits: Regular sustainability audits evaluate the environmental impact of mining operations and ensure compliance with standards. These audits help identify areas for improvement, track progress, and demonstrate accountability to stakeholders.

Reporting and Transparency: Publicly reporting on sustainability efforts and outcomes increases transparency and builds trust with stakeholders. Annual sustainability reports can detail energy usage, carbon emissions, and initiatives taken to reduce environmental impact, showcasing the operation's commitment to sustainability.

Community Engagement and Corporate Social Responsibility Initiatives

Engaging with local communities and participating in corporate social responsibility (CSR) initiatives can enhance the social and environmental impact of mining operations.

Community Engagement:

- **Stakeholder Consultations:** Regular consultations with local communities help address concerns and build positive relationships. Engaging with stakeholders ensures that the community's needs and interests are considered in decision-making processes.
- **Local Investments:** Investing in local infrastructure, such as schools, healthcare facilities, and public services, can benefit the community and improve the mining operation's reputation. These investments demonstrate a commitment to the well-being of the local population.

CSR Initiatives:

- **Environmental Projects:** Supporting local environmental projects, such as reforestation, conservation efforts, and clean water initiatives, can offset the environmental impact of mining. These projects contribute to the overall sustainability of the region.
- **Education and Awareness:** Promoting education and awareness about environmental sustainability within the community fosters a culture of responsibility and stewardship. Initiatives can include workshops, seminars, and partnerships with educational institutions.

Renewable Energy Solutions and Innovations

Adopting renewable energy sources is a critical step towards sustainable mining. Innovations in renewable energy can help miners reduce their carbon footprint and operational costs.

Solar Energy:

- **Setup and Efficiency:** Solar panels can be installed on-site to provide a clean and renewable source of energy. Advances in solar technology have improved efficiency and reduced costs, making it a viable option for mining operations.
- **Cost and ROI:** The initial investment in solar energy infrastructure can be offset by long-term savings on electricity costs and potential incentives or subsidies from governments and energy providers.

Wind Energy:

- **Benefits and Challenges:** Wind turbines can generate significant amounts of electricity, especially in regions with consistent wind patterns. However, the feasibility of wind energy depends on geographic and climatic conditions.

- **Integration:** Combining wind energy with other renewable sources, such as solar, can provide a more stable and reliable energy supply, mitigating the intermittency issues associated with wind power.

Hydro Energy:

- **Sustainability and Scalability:** Hydroelectric power is one of the most sustainable and scalable renewable energy sources. It provides a consistent and reliable energy supply, making it ideal for large-scale mining operations.
- **Case Studies:** Several mining operations in regions like Iceland and Canada successfully use hydroelectric power, demonstrating its feasibility and benefits. These case studies can serve as models for other mining operations looking to adopt hydro energy.

Natural Gas:

- **Cost-Effectiveness and Setup:** Natural gas can be a cost-effective energy source for mining, especially in regions with abundant supplies. It produces fewer emissions compared to coal and oil, making it a cleaner alternative.
- **Environmental Considerations:** While natural gas is cleaner than other fossil fuels, it is still a non-renewable resource and contributes to carbon emissions. Miners should balance its use with renewable energy sources to reduce their overall environmental impact.

Nuclear Power:

- **High Energy Efficiency:** Nuclear power offers high energy efficiency and can provide a stable and reliable energy supply for mining operations. It is capable of generating large amounts of electricity with minimal carbon emissions.
- **Safety and Regulatory Considerations:** The use of nuclear power involves strict safety and regulatory requirements.

Public perception and potential risks must be carefully managed. Despite these challenges, nuclear power remains a viable option for sustainable mining.

- **Feasibility:** While nuclear power can be an effective solution, its feasibility depends on local regulations, infrastructure availability, and public acceptance.

Geothermal Energy:

- **Sustainable Energy Production:** Geothermal energy harnesses the Earth's internal heat, providing a consistent and sustainable energy source. It is particularly viable in regions with significant geothermal activity.
- **Geographic Limitations:** The availability of geothermal energy is limited to specific geographic areas. However, where feasible, it can offer a reliable and low-emission energy solution.
- **Practical Applications in Mining:** Case studies from Iceland and other geothermal-rich regions illustrate successful integration of geothermal energy into mining operations, showcasing its potential as a sustainable energy source.

Considerations for Energy Selection:

Cost and ROI Analysis: Evaluating the cost and return on investment for different energy sources is crucial. Renewable energy options may have higher initial costs but offer long-term savings and stability. Conducting thorough financial analyses helps in making informed decisions.

Environmental Impact: Assessing the environmental impact of each energy source helps in making sustainable choices. Renewable energy sources generally have lower carbon footprints compared to fossil fuels, contributing to a cleaner and more sustainable operation.

Regulatory Compliance: Ensuring compliance with local and international regulations related to energy usage and emissions is essential. Some regions may offer incentives or subsidies for using renewable energy, further enhancing its appeal.

Scalability and Long-Term Sustainability: The scalability and long-term sustainability of energy sources should be considered. Renewable energy sources provide a more sustainable solution for future growth, ensuring the viability of mining operations in the long run.

Demands for Powering Off Mining to Support the Local Grid: During peak energy use, miners may need to reduce their energy consumption to support the local grid. Participating in demand response programs can provide financial incentives and improve grid stability, demonstrating the mining operation's commitment to community and environmental responsibility.

Environmental sustainability and compliance are integral to the future of cryptocurrency mining. By thoroughly assessing their environmental impact, implementing strategies to reduce energy consumption, and adhering to legal frameworks, miners can achieve greater sustainability and regulatory compliance. Engaging with sustainability certifications, audits, and community initiatives further enhances their commitment to environmental responsibility.

Adopting renewable energy solutions and innovations is crucial for reducing the carbon footprint of mining operations. By exploring various renewable energy options and implementing effective energy management strategies, miners can create a more sustainable and profitable future.

In the dynamic world of cryptocurrency mining, staying informed about environmental trends and regulations, and continuously improving sustainability practices, will ensure long-term success and positive contributions to the global ecosystem. Through proactive efforts and responsible practices, the mining industry can balance profitability with environmental stewardship, paving the way for a sustainable and resilient future.

Chapter 21:
Final Thoughts

As we conclude this comprehensive guide to Bitcoin mining, it's important to recap the key points covered, provide encouragement for aspiring miners, and offer some final thoughts on the future of mining. This final chapter aims to consolidate your understanding, inspire you to embark on or continue your mining journey, and provide a clear outlook on what lies ahead in the ever-evolving world of cryptocurrency mining.

Recap of Key Points

Throughout this book, we have explored the multifaceted world of Bitcoin mining, delving into its technical, economic, regulatory, and environmental aspects. Here are the essential takeaways from each chapter:

Fundamentals of Cryptocurrency and Mining:

- **Blockchain Technology:** We began with a deep dive into blockchain technology, understanding its structure, function, and role in the digital economy. Blockchain's decentralized and secure nature forms the backbone of all cryptocurrencies.
- **Cryptocurrency Basics:** We explored what cryptocurrencies are, the concept of decentralization, and how cryptocurrencies operate within the blockchain framework. This foundation is crucial for understanding the intricacies of mining.
- **Wallets and Keys:** The security of cryptocurrency holdings hinges on understanding wallets and keys. We covered the

types of wallets, how to set them up securely, and the importance of private and public keys.

- **Bitcoin Mining Basics:** We detailed the concept of mining, the Proof of Work (PoW) mechanism, and the role miners play in maintaining the blockchain. The chapter also explained mining rewards and transaction fees, crucial for understanding mining economics.
- Evolution of Crypto Mining: The journey of mining from CPUs to advanced ASICs highlighted technological advancements and their impact on mining efficiency and centralization.
- **Choosing Coins to Mine:** While this book focuses on Bitcoin, considering other mineable cryptocurrencies (altcoins) is vital for strategic diversification and maximizing profitability.

Setting Up and Optimizing Your Mining Operation:

- **Mining Rig Setup:** We provided a step-by-step guide to setting up your mining rig, selecting the right hardware, and configuring it for optimal performance.
- **Mining Software:** Choosing and configuring the right mining software is crucial for efficiency. This chapter offered insights into popular mining software options and their configurations.
- **Solo vs. Pool Mining:** We compared solo and pool mining, discussing the advantages and disadvantages of each approach and providing guidance on joining a mining pool.
- **Using Mining Facilities:** The pros and cons of third-party hosting services were explored, along with tips on choosing reliable hosting services.
- **Electrical and Network Setup:** Proper electrical and network configurations are essential for stable operations. This chapter covered the basics of setting up a reliable power supply and network infrastructure.

- **Energy Sources:** We discussed various energy sources for mining, from traditional grid electricity to renewable options like solar, wind, hydro, and geothermal energy. Each energy source's considerations were outlined, including cost, environmental impact, and scalability.

Advanced Mining Strategies and Considerations:

- **Hash Rate Optimization:** Techniques for optimizing hash rate, including overclocking and undervolting, were explored to enhance mining efficiency and profitability.
- **Scaling Operations:** Strategies for scaling up mining activities, managing multiple rigs, and optimizing power and cooling were discussed. Handling increased maintenance and monitoring was also covered.
- **Future Forecast:** We examined trends in cryptocurrency mining, technological advancements, regulatory impacts, and predictions for the future. Emerging technologies like AI, IoT, and blockchain innovations were highlighted for their potential impact on mining.
- **Economic Aspects:** The economic principles behind cryptocurrency, market dynamics, and profitability analysis were detailed. Operational costs, ROI calculations, and hashprice management were key focus areas.
- **Legal Considerations:** Navigating the legal landscape, understanding regulatory risks, and compliance strategies were essential for operating legally and sustainably. Tax implications and staying informed about laws were also covered.
- **Environmental Sustainability:** Assessing environmental impact, reducing energy consumption, and adhering to legal standards were discussed. Engaging in sustainability certifications and audits, community engagement, and adopting renewable energy solutions were highlighted for their importance in achieving environmental responsibility.

Encouragement for Aspiring Miners

Embarking on the journey of Bitcoin mining can be both challenging and rewarding. Here are some words of encouragement for aspiring miners:

Stay Informed and Adaptable: The cryptocurrency mining landscape is dynamic, with continuous technological advancements and regulatory changes. Staying informed about the latest trends, innovations, and regulations is crucial for success. Be adaptable and willing to evolve your strategies as the industry progresses.

Invest in Education: Knowledge is your most valuable asset. Invest time in learning about blockchain technology, mining hardware, software configurations, and market dynamics. This book provides a solid foundation, but ongoing education through forums, courses, and industry news will keep you ahead of the curve.

Start Small and Scale Gradually: Starting with a small, manageable setup allows you to gain hands-on experience without significant financial risk. As you become more confident and knowledgeable, you can gradually scale your operations, leveraging your learnings to optimize efficiency and profitability.

Embrace Innovation: The mining industry is driven by innovation. Embrace new technologies, from advanced cooling solutions to AI-driven optimization tools. Being open to innovation can give you a competitive edge and enhance the sustainability of your operations.

Prioritize Sustainability: Environmental sustainability is not just a regulatory requirement; it is a moral imperative. Prioritize sustainable practices, invest in renewable energy sources, and engage in community and corporate social responsibility initiatives. Sustainable mining practices not only benefit the environment but also improve public perception and long-term viability.

Network and Collaborate: Engage with the mining community through forums, social media, and industry events. Networking with other miners, developers, and industry experts can provide valuable insights, support, and collaboration opportunities. Sharing knowledge and experiences strengthens the community and fosters innovation.

Future Outlook

The future of cryptocurrency mining is both promising and challenging. As the industry matures, miners will face evolving technologies, regulatory landscapes, and market dynamics. Here are some final thoughts on the future outlook:

Technological Advancements: Technological advancements will continue to shape the mining landscape. Innovations in hardware, software, and energy solutions will drive efficiency and profitability. Emerging technologies like quantum computing and AI will bring new opportunities and challenges. Staying at the forefront of these advancements will be crucial for competitive advantage.

Regulatory Developments: The regulatory environment for cryptocurrency mining will become more defined as governments and international bodies develop clearer frameworks. Miners must stay informed about regulatory changes, engage with policymakers, and adopt proactive compliance strategies. Navigating the regulatory landscape effectively will ensure legal and sustainable operations.

Environmental Responsibility: The environmental impact of mining will remain a focal point. Miners will need to balance profitability with sustainability by adopting renewable energy solutions, optimizing energy consumption, and engaging in environmental and community initiatives. Sustainable mining practices will be essential for the industry's long-term viability and public acceptance.

Market Dynamics: Cryptocurrency markets are inherently volatile, with cycles of booms and busts. Miners must develop robust strategies to navigate market fluctuations, optimize profitability, and manage risks. Diversifying mining operations across different cryptocurrencies and staying agile in response to market changes will enhance resilience.

Community and Collaboration: The mining community will continue to play a vital role in the industry's growth and development. Collaboration, knowledge sharing, and collective problem-solving will drive innovation and address common challenges. A strong and connected community will foster a supportive environment for miners at all levels.

Closing Thoughts: As we conclude this journey through the intricate world of Bitcoin mining, it's clear that the industry offers immense potential for those willing to invest in knowledge, innovation, and sustainability. Whether you are an aspiring miner or an experienced professional, the principles and strategies outlined in this book provide a comprehensive guide to navigating the complexities and seizing the opportunities in cryptocurrency mining.

Embrace the challenges, stay committed to continuous learning, and prioritize sustainable practices. The future of mining is bright for those who are informed, adaptable, and responsible. With determination and strategic planning, you can achieve success and contribute to the dynamic and evolving world of cryptocurrency mining.

Appendices

Glossary of Terms

ASIC (Application-Specific Integrated Circuit)

A specialized hardware component designed specifically for mining cryptocurrencies, offering higher efficiency and performance compared to general-purpose hardware like CPUs and GPUs. ASICs are optimized for a single task, making them extremely efficient at processing the specific algorithm they are designed for.

Altcoin

Any cryptocurrency other than Bitcoin. Examples include Ethereum, Litecoin, and Ripple. Altcoins often aim to improve upon the limitations of Bitcoin, offering different features or consensus mechanisms.

Blockchain

A decentralized digital ledger that records transactions across multiple computers in a secure and immutable way. Each block contains a list of transactions and a reference to the previous block, forming a chain.

Block

A collection of transactions bundled together and added to the blockchain. Blocks are verified by miners and must conform to specific rules set by the network.

Block Reward

The reward given to miners for successfully adding a new block to the blockchain, often consisting of newly minted cryptocurrency and transaction fees. Block rewards are a crucial incentive for miners to secure the network.

CPU (Central Processing Unit)

The primary component of a computer that performs most of the processing. Early cryptocurrency mining was done using CPUs before more efficient hardware like GPUs and ASICs became available.

Cryptographic Hash

A function that converts an input (or 'message') into a fixed-length string of bytes, typically a hash code, used to secure data and validate transactions in the blockchain. Common hashing algorithms include SHA-256 and Scrypt.

Decentralization

The distribution of authority and control away from a central entity, such as a single server or organization, across a network of participants. Decentralization enhances security and reduces the risk of single points of failure.

Difficulty Adjustment

A feature of cryptocurrency networks that automatically adjusts the complexity of mining tasks to ensure consistent block creation times, regardless of the number of miners or their combined hash rate. This maintains network stability and security.

Double-Spending

A potential flaw in digital cash schemes where the same single digital token can be spent more than once. Blockchain technology prevents double-spending by ensuring that each transaction is verified and recorded.

Encryption

The process of converting information into a secure format that cannot be read without a key. Encryption ensures the confidentiality and integrity of data, crucial for secure cryptocurrency transactions.

Energy Efficiency

A measure of how effectively a mining rig converts electricity into cryptographic hashes. Higher energy efficiency means lower operational costs and reduced environmental impact.

FPGAs (Field-Programmable Gate Arrays)

Integrated circuits that can be configured by the user after manufacturing to perform specific tasks, including cryptocurrency mining, though they are generally less efficient than ASICs.

GPU (Graphics Processing Unit)

A specialized processor originally designed to render graphics, which is also highly effective for parallel processing tasks like cryptocurrency mining. GPUs are versatile and can mine a variety of cryptocurrencies.

Hash Rate

The speed at which a mining rig can complete cryptographic hashing tasks, typically measured in hashes per second. A higher hash rate increases the chances of successfully mining a block.

Hot Wallet

A cryptocurrency wallet connected to the internet, providing easy access but increased vulnerability to hacking. Hot wallets are used for daily transactions but should not store large amounts of cryptocurrency.

KYC (Know Your Customer)

Regulatory requirements for businesses to verify the identity of their clients to prevent illegal activities such as money laundering. KYC processes are critical for exchanges and other financial services involving cryptocurrencies.

Ledger

A record-keeping system for tracking transactions, balances, and other data. In cryptocurrency, the ledger is maintained as a blockchain.

Mining

The process of validating and adding transactions to the blockchain by solving complex cryptographic puzzles, which secures the network and allows for the creation of new coins. Mining requires significant computational power and energy.

Mining Pool

A group of miners who combine their computational resources to increase the probability of finding a block, sharing the rewards based on contributed hash rate. Mining pools make mining more accessible and profitable for small miners.

Nuclear Power

Energy generated through nuclear reactions, considered for mining due to its high efficiency and low carbon emissions. Nuclear power can provide a stable and large-scale energy source for mining operations.

Overclocking

Increasing the clock rate of a computer's hardware beyond the manufacturer's specifications to boost performance, commonly used in mining rigs to increase hash rates. Overclocking requires careful thermal management to avoid overheating.

Peer-to-Peer (P2P) Network

A decentralized network where each participant (peer) has equal authority and can initiate or complete transactions without relying on a central server. P2P networks are fundamental to blockchain technology.

Private Key

A secure key used to sign transactions and access cryptocurrency funds. Must be kept confidential. The private key allows the owner to prove ownership and control over the funds associated with their public key.

Proof of Work (PoW)

A consensus algorithm used by many cryptocurrencies, including Bitcoin, that requires miners to solve cryptographic puzzles to validate transactions and create new blocks. PoW ensures the security and integrity of the blockchain.

Public Key

A cryptographic key that can be shared publicly and is used to receive cryptocurrency transactions. The public key is derived from the private key and forms the address to which others can send funds.

Renewable Energy

Energy from sources that are naturally replenishing, such as solar, wind, and hydro, increasingly used in mining to reduce environmental impact. Renewable energy sources offer sustainable alternatives to traditional fossil fuels.

ROI (Return on Investment)

A measure of the profitability of an investment, calculated as the net profit divided by the initial investment cost. ROI helps miners evaluate the financial viability of their operations.

Scrypt

A cryptographic algorithm used by some cryptocurrencies, like Litecoin, that requires more memory, making it resistant to ASIC mining. Scrypt is designed to be more accessible to individual miners using general-purpose hardware.

SHA-256

The cryptographic hash function used in Bitcoin mining, producing a fixed-size output from any input size. SHA-256 is known for its security and efficiency, making it ideal for proof-of-work systems.

Smart Contract

A self-executing contract with the terms of the agreement directly written into code, running on a blockchain. Smart contracts automatically enforce and execute agreements without the need for intermediaries.

Soft Fork

A software update that is backward-compatible, meaning that new rules can be added to the blockchain protocol without requiring all nodes to upgrade. Soft forks allow for gradual improvements and updates to the network.

Solo Mining

Mining independently without joining a mining pool, where the miner receives the entire block reward if a block is successfully mined. Solo mining can be more rewarding but also riskier due to lower chances of finding blocks.

Stablecoin

A type of cryptocurrency designed to have a stable value by being pegged to a reserve asset, like fiat currency. Stablecoins provide stability in the volatile cryptocurrency market.

Staking

Participating in the network security and operations of a blockchain by locking up a certain amount of cryptocurrency, often earning rewards in return. Staking is used in proof-of-stake (PoS) and other consensus mechanisms.

Thermal Management

Techniques used to control the temperature of mining hardware, ensuring optimal performance and longevity. Effective thermal management includes cooling solutions like fans, liquid cooling, and immersion cooling.

Transaction Fee

A small fee paid by users to have their transactions processed by miners, which can vary based on network congestion and transaction size. Transaction fees incentivize miners to include transactions in the blockchain.

Undervolting

Reducing the voltage supplied to a mining rig's hardware to decrease power consumption and heat output, often used to improve energy efficiency. Undervolting can extend hardware lifespan and reduce operational costs.

Wallet

A digital tool that allows users to store, send, and receive cryptocurrency, consisting of a pair of cryptographic keys (private and public). Wallets can be hot (connected to the internet) or cold (offline storage).

Whale

A term used to describe individuals or entities that hold a large amount of cryptocurrency, capable of influencing market prices with their transactions. Whales can impact market liquidity and volatility.

Zero-Knowledge Proof

A cryptographic method that allows one party to prove to another that a statement is true without revealing any specific information about the statement. Zero-knowledge proofs enhance privacy and security in blockchain transactions.

Frequently Asked Questions about Bitcoin Mining

What is Bitcoin mining? Bitcoin mining is the process of validating and adding new transactions to the Bitcoin blockchain by solving complex cryptographic puzzles. Miners use specialized hardware to perform these calculations and are rewarded with newly created Bitcoins and transaction fees.

Why is Bitcoin mining important? Bitcoin mining secures the network, verifies transactions, and ensures the integrity of the blockchain. It also introduces new Bitcoins into circulation through block rewards.

What equipment do I need to start mining Bitcoin? To start mining Bitcoin, you need specialized hardware called ASIC (Application-Specific Integrated Circuit) miners, a reliable power supply, internet connection, and mining software.

How much does it cost to start mining Bitcoin? The initial cost of mining Bitcoin can vary widely, depending on the price of ASIC miners, electricity rates, cooling solutions, and other operational expenses. Starting costs can range from a few thousand to several hundred thousand dollars.

What is an ASIC miner? An ASIC miner is a device specifically designed to perform the cryptographic hashing functions required for Bitcoin mining. ASIC miners are much more efficient than general-purpose CPUs or GPUs for this task.

Is Bitcoin mining profitable? Bitcoin mining can be profitable, but profitability depends on several factors, including hardware efficiency, electricity costs, Bitcoin market price, and mining difficulty. It's essential to calculate potential earnings and costs before starting.

How does Bitcoin mining difficulty work? Bitcoin mining difficulty adjusts approximately every two weeks based on the total computational power (hash rate) of the network. This adjustment ensures that new blocks are added to the blockchain roughly every 10 minutes.

What are mining pools, and should I join one? Mining pools are groups of miners who combine their computational power to increase the chances of solving a block. Joining a mining pool provides more consistent rewards, as the pooled effort results in more frequent block discoveries.

What is a block reward? A block reward is the amount of Bitcoin awarded to the miner or mining pool that successfully mines a new block. The reward includes newly minted Bitcoins and transaction fees from the transactions included in the block.

What is the current Bitcoin block reward? The current Bitcoin block reward is 6.25 BTC per block, as of the latest halving event in May 2020. The reward halves approximately every four years in an event known as "halving."

What is Bitcoin halving? Bitcoin halving is an event that occurs approximately every four years, reducing the block reward by 50%. This mechanism limits the total supply of Bitcoin to 21 million coins.

How much electricity does Bitcoin mining consume? Bitcoin mining consumes a significant amount of electricity. The exact consumption depends on the efficiency of the mining hardware and the size of the mining operation. Efficient energy use and renewable energy sources are essential for sustainable mining.

Can I mine Bitcoin on my personal computer? Mining Bitcoin on a personal computer is not feasible due to the high computational power required and the competition from specialized ASIC miners. Personal computers do not provide sufficient hash rate to compete effectively.

What are the environmental impacts of Bitcoin mining? Bitcoin mining's environmental impact includes high energy consumption and carbon emissions, especially when powered by fossil fuels. Using renewable energy sources can mitigate these impacts and promote sustainable mining practices.

What is Proof of Work (PoW)? Proof of Work (PoW) is a consensus algorithm used by Bitcoin and other cryptocurrencies, requiring miners to solve complex mathematical puzzles to validate transactions and create new blocks. PoW ensures network security and prevents double-spending.

What are the tax implications of Bitcoin mining? Bitcoin mining has various tax implications, including income tax on mining rewards and capital gains tax when selling mined Bitcoins. It's essential to keep detailed records and consult with tax professionals for compliance.

How do I join a mining pool? To join a mining pool, you need to choose a reputable pool, create an account, configure your mining hardware and software with the pool's settings, and start mining. Each pool has specific setup instructions available on their websites.

What is solo mining? Solo mining is when a miner attempts to mine Bitcoin independently, without joining a mining pool. While the rewards can be higher, the probability of solving a block is lower, making earnings less consistent.

What is the best mining software? The best mining software depends on your hardware and operating system. Popular options include CGMiner, BFGMiner, and NiceHash. It's essential to choose software that supports your hardware and offers features that meet your needs.

How do I optimize my mining hash rate? Optimizing your mining hash rate involves using efficient hardware, optimizing mining software settings, ensuring proper cooling, and potentially overclocking your hardware. Regular monitoring and maintenance are crucial for sustained performance.

What is overclocking, and is it safe for mining? Overclocking involves increasing the clock rate of your mining hardware beyond the manufacturer's specifications to boost performance. While it can improve hash rate, it also generates more heat and can reduce hardware lifespan if not managed properly.

What is undervolting, and why is it used in mining? Undervolting reduces the voltage supplied to mining hardware, decreasing power consumption and heat output. It improves energy efficiency and can extend hardware lifespan, making it a popular technique among miners.

What are the risks of Bitcoin mining? Risks of Bitcoin mining include fluctuating Bitcoin prices, increasing mining difficulty, high operational costs, regulatory changes, and hardware failures. Conducting thorough research and risk management strategies are essential for successful mining.

Can I use renewable energy for Bitcoin mining? Yes, renewable energy sources like solar, wind, hydro, and geothermal can power Bitcoin mining operations. Using renewable energy reduces environmental impact and can lower operational costs in the long run.

How do I stay updated on Bitcoin mining trends and news? Staying updated on Bitcoin mining trends and news involves following industry blogs, forums, and news websites. Joining mining communities, attending conferences, and subscribing to newsletters can also provide valuable insights and updates.